SYMBOLIC COMPUTATION

Computer Graphics – Systems and Applications

Springer-Series
SYMBOLIC COMPUTATION
Computer Graphics – Systems and Applications

J. L. Encarnação, E. G. Schlechtendahl:
Computer Aided Design. Fundamentals and System Architectures. IX, 346 pages. 183 figs., 1983
Second, revised and extended edition in prep.

G. Enderle, K. Kansy, G. Pfaff:
Computer Graphics Programming. GKS – The Graphics Standard. Second, revised and enlarged edition. XXIII, 651 pages, 100 figs., 1987

J. L. Encarnação, R. Schuster, E. Vöge (eds.):
Product Data Interfaces in CAD/CAM Applications. Design, Implementation and Experiences. IX, 270 pages, 147 figs., 1986

U. Rembold, R. Dillmann (eds.):
Computer-Aided Design and Manufacturing. Methods and Tools. Second, revised and enlarged edition. XIV, 458 pages, 304 figs., 1986

Y. Shirai:
Three-Dimensional Computer Vision. XII, 297 pages, 313 figs., 1987

D. B. Arnold, P. R. Bono:
CGM and CGI. Metafile and Interface Standards for Computer Graphics. XXIII, 279 pages, 103 figs., 1988

J. L. Encarnação, P. C. Lockemann (eds.):
Engineering Databases. XII, 229 pages, 152 figs., 1990

José L. Encarnação
Peter C. Lockemann (Eds.)

Engineering Databases

Connecting Islands of Automation Through Databases

With Contributions by
K.R. Dittrich J.L. Encarnação W. Fischer
W. Gotthard D. Köhler P.C. Lockemann
T. Riedel-Heine M. Ungerer

Springer-Verlag
Berlin Heidelberg New York
London Paris Tokyo Hong Kong

Professor Dr. José L. Encarnação
TH Darmstadt, FG GRIS
Wilhelminenstraße 7, D-6100 Darmstadt, FRG

Professor Dr. Peter C. Lockemann
Fakultät für Informatik, Universität Karlsruhe
Postfach 6980, D-7500 Karlsruhe 1, FRG

With 152 Figures

CR Classification (1987): H.2, H.4, D.2, B.7

ISBN-13:978-3-642-64859-5 e-ISBN-13:978-3-642-61508-5
DOI:10.1007/978-3-642-61508-5

Library of Congress Cataloging-in-Publication Data.
Engineering data bases / José Encarnação, Peter Lockemann, eds. p. cm.-(Symbolic computation. Computer graphics-systems and applications) Includes bibliographical references.
ISBN-13:978-3-642-64859-5
1. Engineering-Data bases. I. Encarnação, José Luis. II. Lockemann, P. C. III. Series.
TA345.E543 1990 025.06'62-dc20 90-9590 CIP

Softcover reprint of the hardcover 1st edition 1990

2145/3140(3011)-543210 - Printed on acid-free paper

Introduction

Automation is nothing new to industry. It has a long tradition on the factory floor, where its constant objective has been to increase the productivity of manufacturing processes. Only with the advent of computers could the focus of automation widen to include administrative and information-handling tasks. More recently, automation has been extended to the more intellectual tasks of production planning and control, material and resource planning, engineering design, and quality control. New challenges arise in the form of flexible manufacturing, assembly automation, and automated floor vehicles, to name just a few.

The sheer complexity of the problems as well as the state of the art has led scientists and engineers to concentrate on issues that could easily be isolated. For example, it was much simpler to build CAD systems whose sole objective was to ease the task of drawing, rather than to worry at the same time about how the design results could be interfaced with the manufacturing or assembly processes. It was less problematic to gather statistics from quality control and to print reports than to react immediately to first hints of irregularities by interfacing with the designers or manufacturing control, or, even better, by automatically diagnosing the causes from the design and planning data. A heavy – though perhaps unavoidable – price must today be paid whenever one tries to assemble these isolated solutions into a larger, integrated system.

Integration is a matter of cooperation between systems, each dedicated to a particular set of tasks in the industrial process. Cooperation is based on communication. Communication requires a technical infrastructure of hardware and software facilities. To provide these is the objective of informatics. But communication also requires rules – or, to use a modern term, protocols – that enable individual systems to talk to one another. These rules are not entirely technical in nature, but have also to do with the mission, organization and philosophy of the enterprise. To set up effective communication, to integrate the isolated islands of automation, is a challenge that must equally be shared by managers, supervisors, automation engineers, communication engineers, and computer scientists.

Communication takes place in space and in time. Communication in space is concerned with the speedy delivery of information from one geographic location to another. By contrast, communication in time is concerned with relaying information from one point in time to another, necessitating storage of the information for definite or indefinite periods of time. Databases are the tools that provide the required support.

This book deals with technical communication in time as one of the means for facilitating integration of what until now have been islands of automation. Hence, its subject is database technology as it applies to technical processes, i.e., those in planning, design and manufacturing. The emphasis will be on both technology and protocols, acquainting the reader with the technical perspectives of database systems and the user interfaces and standards to be observed.

Database systems today are a well-established technology that has proven its worth in tens of thousands of installations, ranging from mainframe systems with databases of hundreds of gigabytes, all the way to systems on workstations and personal computers. In the course of more than 20 years of development, database systems have matured to satisfy practically all needs regarding data-structuring facilities, reliability, simultaneity of access by multiple users from multiple locations, and performance. Or to be more candid, the needs have been satisfied provided they arose from typical administrative and business applications.

The question that this book addresses is this: Is today's database technology mature enough also to serve the needs of engineering applications and industrial automation? Those at the forefront of database research tend to answer the question in the negative, and in fact devote much of their talent to coming up with new techniques to satisfy the new needs. However, even if they are correct, today's engineering community is forced to live with database system products that reflect the older state of the art. Should they wait until new products become available? The purpose of this book is to demonstrate that even though solutions based on today's technology are less than perfect, they nonetheless provide solutions to current pressing problems. Consequently, the book will mainly cover the current database technology and its applications. On the other hand, it will not completely ignore some of the more promising techniques that seem to be evolving from scientific research, in order to prepare the reader for the opportunities of several years hence.

To whom do we address the book? As mentioned above, integration is the concern of persons of many diverse backgrounds and responsibilities. The book sets out to foster cooperation between these, to build a bridge and improve understanding between managers, engineers, and database technicians. Intended readers are middle management and engineers from industry who deal with automation both as users and vendors, consultants to such industry, vendors of database systems, and lecturers and students in educational institutions that provide training in industrial automation. The book requires no special background in informatics, but expects the reader to have a general understanding of and feeling for technical subjects.

With this clientele in mind, the book avoids excessive technical detail and instead occupies a middle ground between database technology, general issues of database utilization, application of database systems to practical engineering situations, and their impact on the industrial and organizational environment. No attempt is made to cover each and all topics, rather the goal is to raise the awareness of the main issues that arise in connection with database systems as an integration tool. We can thus afford to treat the more recent developments fairly lightly, and only to the extent that they contribute towards an understanding of how to plan for the future.

For cursory reading, each of the four chapters may be considered self-contained and glanced over independently of the others. For in-depth reading we recommend covering the chapters in order. Chapter 1 addresses the need for a comprehensive database solution in an engineering environment primarily from a management perspective. It identifies the integration problems in automation systems, explains how databases may help to overcome them, and outlines the database system architecture needed to render this help. The chapter closes with a discussion of the organizational consequences of introducing databases as an integration tool.

Chapter 2 lays the technical foundations by giving an introduction to database systems. The main emphasis is on – as we termed it above – protocols, i.e., the interface through which users and programmers communicate with database systems. Hence, extensive use is made of examples. Implementation topics are covered only to the extent that they help in understanding the functional capabilities of database systems. Major subjects treated are data organization and manipulation, the integrity issues of consistency and concurrency control, and of recovery, transaction management, and performance control. These are mainly dealt with in terms of existing database technology. The last part of the chapter, however, explores in more detail where weaknesses of the present technology are perceived, and a few brief examples are given of new technical trends.

A database system is nothing more than a tool. Chapter 3 describes how to use the tool in a methodical way in engineering applications. Main topics covered are the design of the database schema, i.e., of the database structure; the management of versions, an important issue for design environments; the generation, acquisition and archiving of data; data distribution and communication; standards for data interchange as a means for communication in open systems; and linguistic means for comfortable database access from application programs. Many of these are not database topics per se, but they are critical to the success of database utilization.

Chapter 4 is devoted to case studies from two large projects that include integrated databases as an essential component of the technical infrastructure. Both are derived from cooperation between science and industry in the Federal Republic of Germany and concern design environments: the first an environment for VLSI design, the second an environment for the development of real-time systems and software. The approach taken for these case studies is the same: In a top-to-bottom manner, first the problems to be solved are introduced, then the information to be communicated is identified, an appropriate database schema is derived, and finally examples of application programming for the databases are given.

A book that claims to foster cooperation between persons from different backgrounds should be written by authors who themselves come from different backgrounds and had to learn to cooperate. Fortunately, this has been the case for this book. Some of the authors are scientists developing new database technology for engineering applications, others are engineers struggling in their everyday work with the lack of integration between engineering tools and shortcomings of present-day database technology, and still others are scientists applying databases to problems in graphics and to making comfortable graphical user interfaces for design environments. All of the authors have extensive experience in industrial projects. The editors are extremely grateful to the authors for giving their precious spare time to this book. They also thank Springer-Verlag for its customary reliable support.

The editors have high hopes that this book will achieve its goal of fostering cooperation between all persons engaged in issues of industrial automation, and in doing so will advance the techniques and solutions for integrating today's islands of automation.

Darmstadt/Karlsruhe J.L. Encarnação
December 1989 P.C. Lockemann

Contents

1 Significance of Engineering Databases

1.1 The Situation

1.1.1 Evolution of Engineering/Scientific Data Processing (CAx)

What has often been overlooked is that scientific and technical problems spawned the invention and development of digital computers to aid in the search for numerical solutions to these problems, i.e., to facilitate the execution of complex calculations. It was not only Konrad Zuse's idea to work on the development of computers to improve computer routines; in fact, it played a dominant role with all American pioneers of the 1940s. In the 1950s universities, public and industrial research centres, and the armed forces were the main purchasers of the first generation of engineering-scientific mainframes, and it was only towards the end of the decade that it became clear that the new computers could be used not only for numerical calculations, but also for a wider range of tasks, e.g., for non-numeric applications.

The wide use of computers in industry began in the 1960s, especially in large companies where large and very expensive mainframes – as they were considered at the time – were installed in computer centres. Data was entered via punched cards (batch mode), the programming language of the engineers was Fortran, and the required numeric data files were chiefly stored on magnetic tapes, and later on magnetic disks.

In the 1960s and 1970s these mainframes, based on efficient operating systems, were used to create large, technical user software packages for a wide range of engineering tasks. This meant that the development engineer, on the one hand, no longer had to spend time programming himself and, on the other hand, could incorporate complicated computing methods, such as finite element analysis, in his daily tasks. Activities in this field also began at MIT in 1960 and this led to the development of the CAE system ICES (Integrated Civil Engineering System) and also of the first CAD/CAM system APT (**A** **P**rograming **T**ool for NC machines).

In the meantime, the situation for the different CAx appplication areas, or "islands", has greatly changed with regard to software and hardware, both from a quality and quantity point of view. A glance at US, European and Far East production plants shows that rapid changes in computer and communication technology are changing the current industrial scene. Even though at present the big industries are without doubt the pacemakers, particularly on the fiercely contested world markets for technical mass products, small-scale industries have also recognized the signs of the automated-engineering revolution and are making changes to be able to meet the requirements of the 'plant of the future'.

This boom was triggered off by rapid developments in hardware technology. The significant factors were enormous increases in computer performance, both in processing speed, and in internal and external storage capacity.

In addition, there have been developments in input and output peripherals which, on the one hand, are characterized by high-precision raster graphics terminals and workstations, and also by devices adapted to them, i.e., the conventional input devices that complement the keyboard such as the mouse, and on the other hand, by output devices which are moderately priced and provide high quality printing results, such as the latest laser printers.

One of the newest hardware developments is reflected by the progress made in the field of local and supraregional networks. All these developments have yielded on-the-spot efficient data processing equipment, for example, the portable "lap top" with a performance that in the 1970s was supplied by a commercial computer, and the workstation, which even exceeds the performance of early mainframes.

Thus, in engineering/scientific data processing, we find ourselves in the middle of a change-over from central batch processing to interactive graphic decentralized operations (with workstations and colour-shaded images).

On the other hand, today an engineering/scientific-system is no longer a stand-alone discipline but is part of an entire network of information processing activities, among other CAx islands, mainly CAE, CAD, CAPP and CAM.

The new demands posed by CAx, and the vast potential of the latest developments in hardware and software techniques, present new challenges to software development.

Software for this kind of computer is nearly always dialogue-oriented, and if necessary interactive graphic. The old division into data processing and word processing will disappear, and will be replaced by something which is today usually referred to as knowledge-based data processing. We also need a new software technology, because the exploration and exploitation of the new alternatives inevitably require a more "experimental" method of developing software products than classical software technologies tend to use.

The challenge is also educational because of the serious lack of specialists. The technical universities and colleges should do more to produce graduates with state-of-the-art knowledge, and should emphasize in their research the conversion of computer science knowledge into practical engineering applications.

1.1.2 Oceans of Information

Within a few years, the acronym CIM (Computer Integrated Manufacturing) has become accepted as the collective term for the whole of product-related

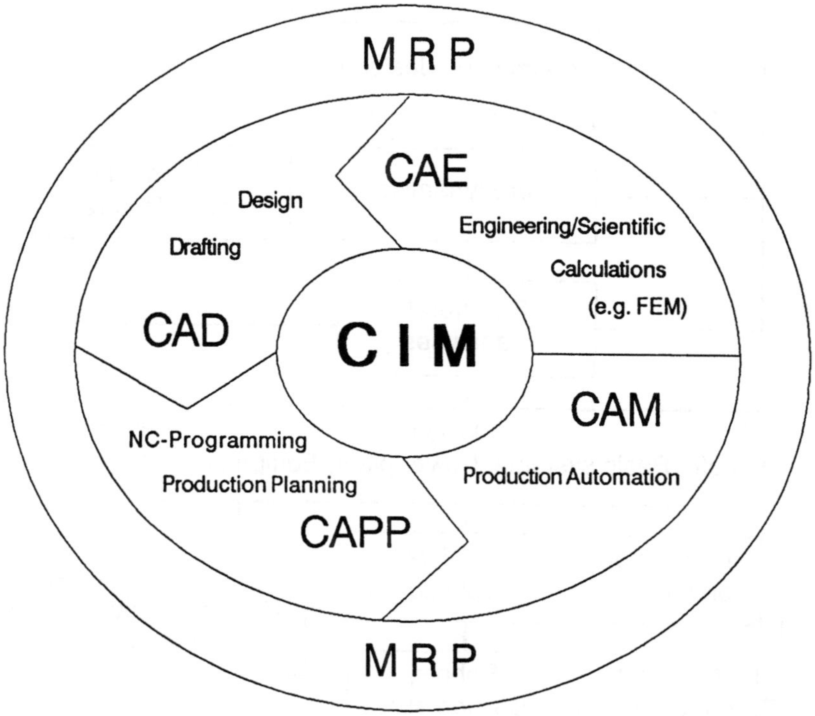

CAE = Computer Aided Engineering
CAD = Computer Aided Design
CAM = Computer Aided Manufacturing
CAPP = Computer Aided Process Planning
MRP = Material Resource Planning

Fig. 1.1. CIM = (CAD + CAE + CAM + CAPP) * MRP

(i.e., technical and technical/ administrative) information processing (see Fig. 1.1). CIM means that a complete and comprehensive computer-aided information and processing flow of all the development, planning and production data from the entire company is established on the basis of a common/centralized computer-maintained product model.

A good, up-to-date, intricate and extensive information flow of all operations is evidently necessary to achieve this. Such an increasing quantity of information can no longer be dealt with by hand. The instrument that is required here for CIM realization is comprehensive, computerized information processing which is based on traditional data processing.

The quality and quantity of this information processing depends on the type of product. Equally, CIM also calls for adequate specialized and organizational competence in the company to be able to function at all, and to achieve the intended effects.

The generation, processing, and transfer of information during the development and life-cycle of a technical product is becoming increasingly

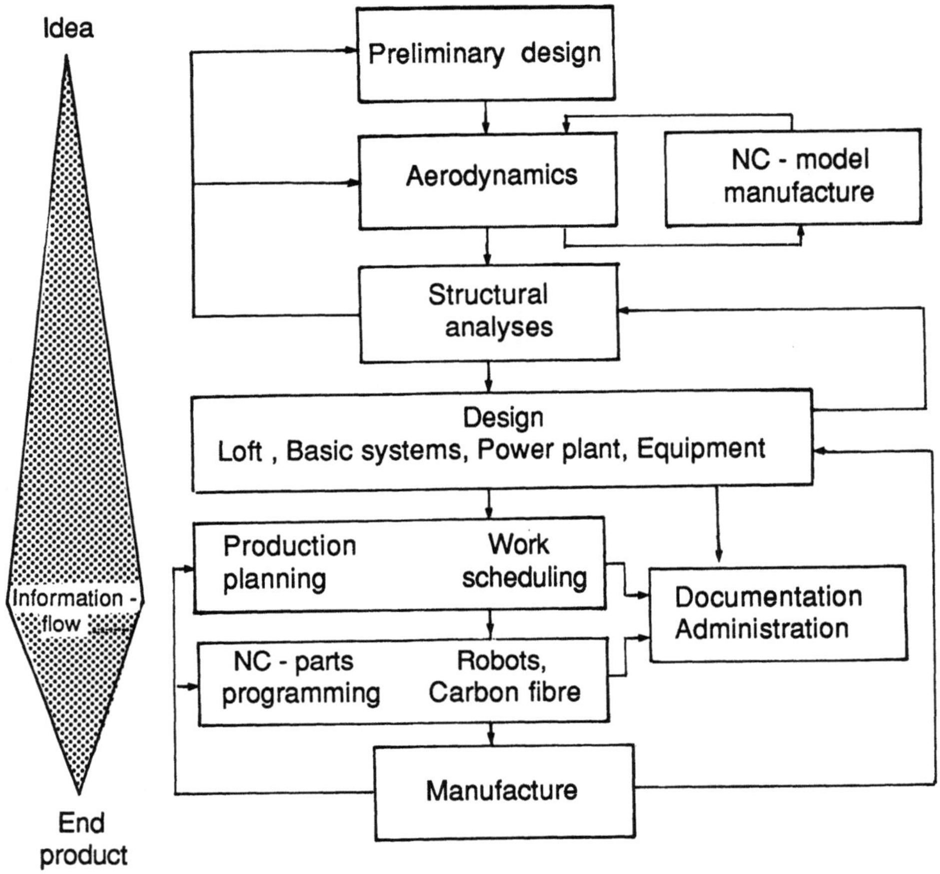

Fig. 1.2. Information flow in aircraft and car construction

intricate and expensive (e.g., in aircraft development and production due to fiber composites technology). On the other hand, the times for development and production, and the service life of these products are becoming steadily shorter.

The volume of information, starting with the conception of an idea, and continuing to production, is also continually increasing, both in quantity and quality. One significant influence on product optimization and quality is, of course, the flow of information between specialist departments. As far as possible, this should be in the form of a control loop: it should be authentic, without delay, and with constant feedback. A good example of this is the information network illustrated in Fig. 1.2.

In addition, the technical-administrative tasks are to be regarded as activities accompanying the actual creation of the product – affiliated with Material Resource Planning (MRP) – (e.g., bills of materials, logistics, configuration management, documentation) which also have to be integrated in the information model of a company.

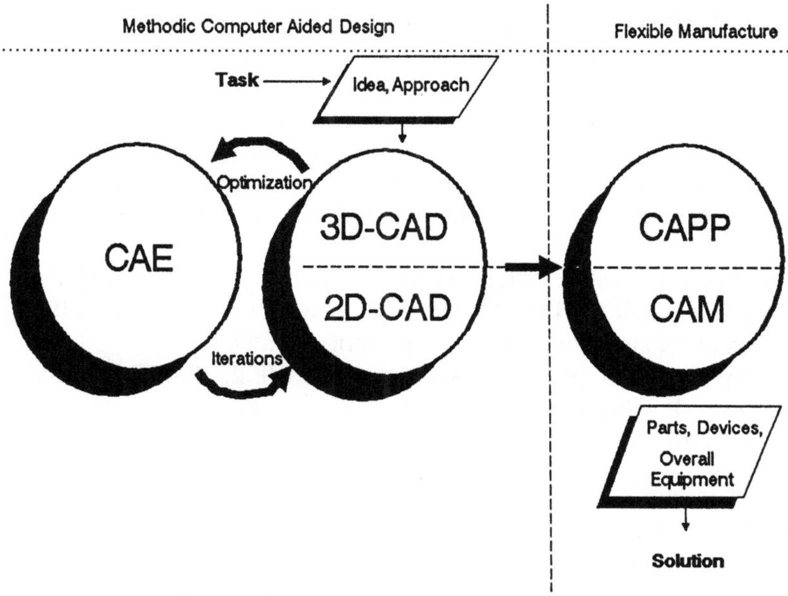

Fig. 1.3. Integrated CAx operations/process chain

The general CIM process chain, i.e., the engineering/scientific part with its integrated CAx operations (see Fig. 1.3), provides a clearer insight into the CIM connections and associations. Here, strong interactions (product-related, of course) between CAD and CAE are apparent. This is nothing but the conventional virtue of the engineer, that is, methodical design; however, now with computer support. This means that the engineer's work is still of great importance.

Generally, it can be said that today there are already very good and self-computing partial solutions available in the industry, predominantly in the CAD area. Considerable rationalization is possible with special solutions; thus, a productivity increase of factor 4 can be achieved, for example, in the graphic-interactive generation of drawings (2D-CAD).

However, the computer-aided systems currently in use are of a more or less singular or autonomous character (insular solutions), and are furthermore for the most part functional, yet also in need of improvement from a data processing point of view (e.g., problems with large amounts of data).

In addition, their user-friendliness often leaves much to be desired (communications philosophy). Here, a difference has to be made between the non-experienced and the experienced user (conversant with EDP) and also, of course, between systems with varying levels of performance (e.g., CAD-2D or -3D).

However, generally speaking it should be noted here that expectations rise as one gains experience. Thus, it will be a long time before one is completely satisfied, that is, before all the requirements are met.

If one takes a look at the various CAx areas, a comparison with CAD and CAM shows that the NC systems are inferior. Firstly, there are insufficient functions to cover the production requirements, and secondly, there is not yet any real graphic-interactive communication.

Graphic-interactive programming of parts using the conversational mode on the graphic terminal yields, as with CAD, an increase in productivity because here, the application program is executed immediately after being input (Code), and is immediately displayed (Show) on the graphic screen. This enables the part programmer to control his results directly and to correct any errors straight away.

CAE still mainly comprises batch-oriented systems (e.g., NASTRAN). However, it is very difficult to change such large program systems to on-line, both from a design and from a data processing point of view.

For this reason, so-called pre- and post-processors (e.g., PATRAN, SUPERTAB) have been developed to try to offset these disadvantages. However, for two reasons this can only be a temporary solution:

1. The engineer handling mathematical applications is "purely the end user", he has to accept the pre-/post-processor system as it is.
2. The data structures and models of the current pre-/post-processors are too weak and inflexible for CAE integration into the CIM network.

With regard to production control with CAM, there are already very good solutions available in the direction of the automated factory, for example programmable controllers, process and master computers, and robots and their coordinated control.

Material Resource Planning systems are the most widely used. The change-over from batch to on-line processing (even real-time to some extent) is in full swing.

The objective here is to enable more accurate machine capacity and manpower planning, and to achieve shorter throughput times, less idle time and downtime, and to be able to give an overall status report of the production situation with up-to-date data regarding the production progress of all orders and the associated and necessary resources.

So we see large and endless oceans of information but without an integrated computerized and controlled information flow and processing. There are only some local and isolated solutions around the different CAx islands.

In other words the efficiency, quality and economy of current CIM-solutions are still extremely unsatisfactory. There is a very large gap between the geometric-technical CIM sector (CAD/NC/CAM) and the technical-administrative CIM sector (CAP, CAPP). Here, emphasis is placed on the functional extension of existing systems, and the improvement of their data processing features ("open systems" with suitable interfaces) in order to accelerate the integration process. Separate optimization efforts on both sides are today by no means sufficient. Economic production automation with overall optimization cannot be achieved without machine-aided, on-line, and integrated administration of the CAD/NC/CAM complex. The parts geometries originate in the CAD field, they are programmed via NC machines and then manufacture is carried out in the CAM field. CAPP is intendend to back up and coordinate the technical subareas (e.g., bills of materials, but also the consequent archiving of geometries and NC-parts programs) and furthermore manage the entire area of production planning and control in the CAM field. This is where there is by far the most potential for rationalization.

1.1.3 Bridging the Islands

Now the big problem is to bridge over the different engineering CAx islands, and also to link them to the administrative areas (MRP). Such a fully computerized information flow is what CIM really means. The ability to bridge depends upon the quality of the information. So the basic need is the portability of information. Portability means the ability to be carried or transferred. Software is "carried" by man (the creator of the contents), physical data carriers (storage media, tapes, etc.) and processing devices (processors, etc.)

The portability of software requires the portability of programs, data (often, or in fact particularly, of old data with regard to release changes), and also of interfaces; i.e., the user ought to have the same user interface when using a different application (portable user interface).

Thus, it is necessary to clearly differentiate software with regard to programs, data and interfaces. In the past the data structures were often somewhat neglected and melted within the programs.

Today, with increasingly complex systems (both in quality and quantity), a more data-oriented concept is absolutely essential. Instruments such as the data dictionary and data management are covered by this, and should definitely be included in the overall information analysis procedure of every large-scale project.

Thus, portability calls for abstraction.

In order to gain portability of programs the ICES language component ICETRAN has been defined, modernized, and extended from its early beginnings. The language volume now contains portable FORTRAN 77 (average of all target system compilers FORTRAN 77), essentially extended to include data structures similar to those of Pascal, with the objective of achieving greater data abstraction. This is a great advantage for defining geometry elements. For the following reasons, we feel that from a current point of view this is the best choice of a higher programming language for the development of software in the technical-scientific field: in FORTRAN 4 and FORTRAN 77 there are already a large number of modules and software available (in particular in the field of complicated mathematical algorithms or computation processes), and FORTRAN is used by the engineer for mathematical applications, i.e., FORTRAN formats are a form of standard for integration in the CAE-CAD field.

For the future, the following approach will bring the best combination of flexibility, efficiency, portability, and performance in engineering/scientific software development and production: merging and enhancing old and new systems by using FORTRAN, PASCAL and C (UNIX) assisted by C or UNIX-based dedicated higher languages (e.g., the language PROLOG for artificial intelligence).

Portability of data in the engineering/scientific field remains a problem. The data exchange between existing CAE, CAD and NC-systems is generally very complicated because of different data model structures, especially with respect to topology and geometry representation (Bezier, Coon, B-Splines, NURBS, etc.). One can verify that very easily by thinking of the different possibilities for defining a circle, for example.

Within integration level 1 (see Fig. 1.4) there are already some good point-to-point connections in use (e.g., CADAM-CATIA-2D), but the general approach must be the bus-concept (integration level 2). The two advantages are, first, interface standards (like CLDATA, IGES, VDAFS, SET, STEP, etc.) and second, a reduction of necessary interfaces (see Fig. 1.4) with the consequence of less software effort. The improvement of such standards should finally bring completeness (all geometrical and technical entities can be handled) and efficiency (the interfacing performance is reasonable).

Level of Integration	Number of Interfaces (Pre- and Post-)
1: point to point	N * (N - 1)
2: BUS concept	2 * N
3: universal models	0

Fig. 1.4. The three levels of integration

The ideal integration level 3 would have provided (future) CAx systems with a universal model structure. Such systems would be able to understand everything. A pragmatic solution will be brought by future expert systems which will simulate such features also for conservative CAx systems. The general approach will be object-oriented programming, because abstraction means nothing else.

1.2 The Solution

1.2.1 Integration Through Databases

After covering (solving) the aspects of information quality (bridging the CAx islands by proper interfaces), there still remains the second large problem, which is the quantity of information. There are two reasons why this is rapidly increasing: on the one hand, new applications continually arise alongside a broader use of the existing ones; on the other hand, the integration idea causes a nearly overwhelming demand for data storage, retrieval and administration.

The database problem was born in the 1960s with the installation of the first mainframes for commercial applications. So big data volumes arose, and to begin with there was organized data collection, followed by file management, and finally database management systems (DBMS). The bill of materials problem (hierarchical models) played a dominant role here.

As it is commonly known today, a database management system carries out the organization of data storage and data management for a whole application area. This provides three major advantages:

1. The organization of data does not necessitate specific programming for the application systems.
2. The application is easier to extend.
3. The application data can also be made accessible to other application systems (via standard interfaces).

These three advantages are of a complementary nature: the high degree of data independence, i.e., a strict division of data from the application programs, creates neutralization between data and their management on the one hand, and the programs on the other hand. That is, the usual program modifications due to reorganization of the data, and vice versa, can be avoided. This distinct separation of application software and system software (this is where database management systems fit in) is what really leads to clear and efficient solutions for larger systems. The same applies to the multiple-usability of data in global systems.

Therefore, database management systems can be regarded as being the realization of the idea behind integrated data processing. Their implementation is an indispensable prerequisite for modern information systems (integrated information processing).

Great progress has been made in the development of classical (commercial) database management systems. In addition to the sophisticated hierarchical and network systems, relational systems are currently also of great importance. Their significance is steadily increasing due to their powerful query and operating languages, and their simplicity with regard to data representation. The user can specify the desired information using attributes; no access features or sophisticated search programs are required. If consistency rules are formulated, data values and data relationships can easily be checked.

The reasons for the success of the relational model are, firstly, the use of ordinary (flat) tables (or relations) in data description, and secondly, the fact that the relational model is set-oriented, i.e., the tables form the basis for queries and processing operations. Similarly, the results of a query or processing operation form a table characterized by the given attributes. The standard SQL is now the generally accepted language for descriptive queries and processing operations (in contrast with the otherwise common procedural language). The third major advantage is that the relational model has a unique formal mathematical character, i.e., it theoretically offers the highest possible degree of flexibility. This means the normalization theory is supported in its entirety, which is necessary for the consistent and semantically well-defined design of a database management system.

The power, flexibility, and sophistication of these relational systems do, of course, have a price, which as we know is particularly high and (still) exorbitantly expensive. The term "engineering database management systems" which recently came into use reveals two things: first, the quality is different from that in classical database management systems (designed for commercial and administrative use); and second, databases have been dreadfully neglected in engineering/scientific data processing.

The question we should ask is whether classical database management systems are the answer to engineering needs. If so, how are they best put to use? If not, what should take their place?

In the 1960s and 1970s (and even today), large user software packages in the various fields of engineering were created on mainframes with efficient operating systems as the basis. This meant, on the one hand, that the designer was relieved of time-consuming programming operations and, on the other hand, that complicated calculation methods such as finite element analysis could to a large extent be incorporated in the routine tasks of engineers. The dilemma over databases arose because to begin with, a single file management system was adequate to read in and write out simply structured data. For too long, the developers of engineering/scientific systems believed they could solve the data problem on their own (of course implicitly in application systems).

Today, we realize that in fact redevelopment is required in this sector with regard to integrated data processing.

The worst symptoms are problems with large datasets, data consistency, and above all data exchange.

1.2.2 CIE: Computer Integrated Enterprise

Today EDP (Electronic Data Processing) is replaced by Information Processing (IP), because now we are aware that information is the fourth and most important resource in enterprises. The strategic main targets of integrated information processing are:

1. Increase in productivity, firstly by improving the individual operating processes, and secondly by their computerized integration.
2. Reduction in run times for the entire product-creation chain (design-manufacture-support) by means of an integrated, machine-aided flow of information.
3. Improvement in quality of products, firstly, by better machine-aided optimization features, and secondly, by better data consistency of higher data quality.

Future information processing will have to serve the complete information needs within an enterprise. This is called CIE (Computer Integrated Enter-

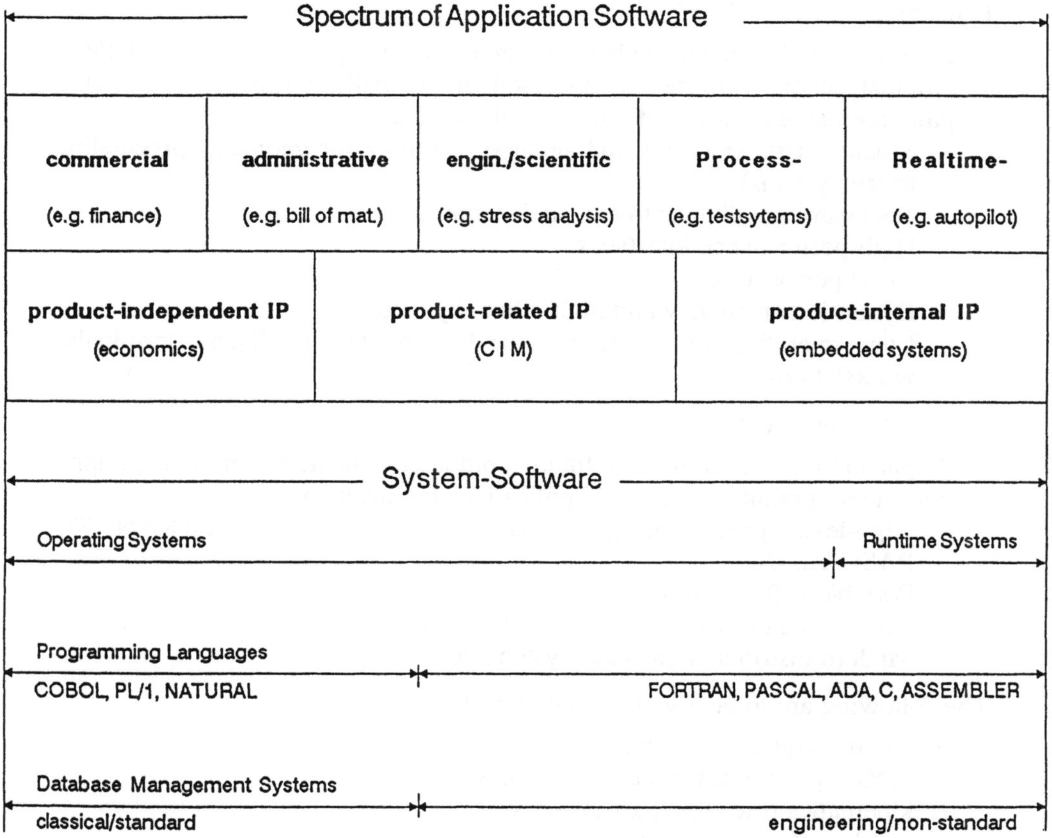

Fig. 1.5. Computer Integrated Enterprise

prise). A rough overview is shown in Fig. 1.5. The area of CIM will be emphasized below.

The CIM strategy for the future means nothing other than the computer-aided compilation of the CAx subareas into an integrated, efficient overall CIM system with a promising future. The CIM systems of the future will have the following features:

Integrated:

- Single source data with automated transfer and further processing which is
  - Automatic (e.g., optimization of cutter path or application of expert systems), and
  - Interactive (e.g., 2D graphics, or extended to 3D)
- Rapid and automated information flow:
  - Development -> transfer to production -> production
- Data quality and quantity:
  - Exactness and clarity of the information in completely automated representation (consistent geometry models with unlimited size).

Efficiency:

- Economic and sensible realization (of CIM) and production (with CIM): Gradual action with the incorporation of internally existing or externally purchased subsystems, and with the following aspects:
  - Modern data structures and suitable operators for geometry processing (open systems),
  - Necessary possibilites to extend functions,
  - High-performance interfaces,
  - Good performance,
  - Dynamic consistency and high reliability, and
  - Low operating costs, decentralized computer, intelligent terminals, workstations.

Good future prospects:

- Application to different and future computers, including graphic peripherals, and possibility to extend applications by means of:
  - High-level programming languages (e.g., portable FORTRAN 77, PASCAL, C),
  - Portable software tools,
  - Universal data structures and interfaces, and
  - Modern distributed database systems for CIM.

The following are to be given special consideration:

- Information quality of data:
  - 2D/3D geometries (incl. composites),
  - kinematics (movable parts, robotics),
  - topology (= associations),
  - physics (e.g., material properties), and
  - administrative data (e.g., bill of materials in associations).
- Information quantity of data:
  - sufficient file and model sizes due to dynamic storage management and transfer technologies, and
  - uniform, flexible information model that covers all products, components (simple and complex) for all qualities of information.
- Data protection and security (as previously):
  - control/inspection of data transfer (e.g., graphic-interactively),
  - protection against hardware and software failures (back up, restart), and
  - management of accesses and data transfers.
- User-tolerated response time behaviour in
  - processing, and
  - transfer.

Two things have to be done on the road towards CIM:

1. **The CIM Analysis.** Do not lay down high requirements for the CIM system; instead, clearly analyze the operational sequences: who needs what and when and in which form?

The necessary methods or instruments, e.g., conceptual scheme, canonical synthesis, data and information management (including EDP-based data dictionaries) are well known today. This course of action is a prerequisite for the successful handling of the whole of today's information processing: in its entirety (integrated), stably (logically consistent), and with good prospects for the future (with long life, but capable of extension for new requirements).

2. **The CIM Synthesis.** In addition to already existing hardware and software, foreseeable developments in these areas must also be incorporated in the concept (prospects for future). This includes new terminals or workstations, or the relational, non-standard, and distributed database systems with enhanced SQL interface which we feel could greatly enhance the overall CIM solution.

The complexity of such an overlapping CIM concept and also the increase in the actual status means there can be no general TOP-DOWN approach for the realization. In the realization phase the pragmatic BOTTOM-UP procedure takes the technical, time, financial, and organizational aspects better into account. Existing and planned insular solutions must be gradually integrated in a uniform overall system with a uniform data basis. The continuing integration of corporate functions offers the chance to continue development at an increasingly high productivity level. The higher integrity level comprises an effective level of the information and data flow and liberates synergy effects.

For these reasons, it is already obvious that CIM will (have to) be a distributed system with local and central parts (thus the CAD designer/engineer will work solely at the local workstation and will only occasionally exchange drawings and/or models with the central database). This of course has to take place with the incorporation of the entire specialized knowledge from individual areas, and all the retroactive effects and structural changes that this involves.

The constant advances and integration of CIM within the framework of overall information processing, and the equally rapid increase in the volume of applications makes it all the more important to have an up-to-date and consistent database. The handling of the various data qualities and quantities and their interfaces/associations (topology, associations, relations) must first be systematically dealt with in a (logical) CIM product model.

Due to volume and complexity, the CIM production model must exhibit different information levels and partial views, so-called subschemata. This overall model is preferably to be EDP-based and maintained, so that it can later be used automatically and reliably in the CIM application systems.

A flexible, i.e., programmable interface is to be defined and realized for data exchange between the various distributed data-bases and the central computer. Geometry data structures and models are under discussion here, because they are considerably more complicated than comparable commercially available models.

1.3 The Architecture of an Engineering Database System

1.3.1 Conceptual Aspects

By restricting the discussion primarily to the CIM area we exclude so-called "embedded systems", these being process data systems and primarily real-time systems (Fig. 1.5). Admittedly, these kinds of systems are an extremely important component of intelligent products; they require reproducible reaction times and this in turn calls for a minimum system overhead and defined data space. Conventional database management systems cannot even be considered for this kind of application, so that these issues are still being dealt with in the language environment.

Unfortunately, even for the CIM applications, technical database management systems – in the science and research sectors these are also classified as non-standard database management systems – with regard to their readiness for production and their operational capability, are a very long way behind classical (standard) database management systems.

There are a number of general conceptual formulations that apply to all database management systems, and thus also to technical databases:

- Integration with single-source data generation and acquisition, computerized transfer and processing;
- Flexibility provided by open interfaces and data structures, and the possibility of extending data models and functions;
- Efficiency by step-by-step procedure, adequate performance and economical operating costs; and
- Good future prospects due to portable software and interfaces (SQL).

However, the following differences between the tasks of commercial and technical systems are of great importance. At the same time, this shows that technical systems are more general, whereas the classical commercial systems are really for particular cases.

Commercial applications contain simple data records with easily structured data; in technical applications, on the other hand, we find a heterogeneous quantity of inhomogeneous data records, and data with complicated structures. Similarly, in the former case the types of data solely contain formatted data, such as scales or chains of characters of fixed length, in the latter case there are multi-dimensionally formatted data like text, illustrations, etc.

A transaction in a commercial system contains only a few data records, the length of the transaction is short, and there is concurrent consistency and recovery. In technical systems there are many different data records and the transaction time is often long (local single user processing period). According to DBMS language rules, this is what is known as a lengthy transaction. This means that consistency and recovery have to be solved in different ways. In archiving, as well, the technical systems have multiple, associated representations, and a version check is also required, whereas in commercial systems simpler representations and coverage of the generation principle are sufficient.

An important practical example can illustrate this: the designer or engineer working at a workstation in the local area network is a single user. He has no concurrency problems except for transmission. The sophisticated multiple write access feature is completely unnecessary in local incore modeling. As a consequence, the general, integrating system must be a distributed non-standard DBMS with central and local parts.

The well-known advantages of a distributed Database Management System are:

- Load distribution according to computer capacity,
- Storage of data with (most important) user,
- Avoidance of unnecessary multiple storage and resultant inconsistency,
- Better use of resources,
- Adaptation to functional and organizational structures,
- High availability, and
- Improved response times.

The following requirements must, of course, be fulfilled:

- Portability of programs,
- Transparency with regard to localization and possible redistribution of data,
- Distribution of identical data among various nodes,
- Controlled data redundancy,
- Coordination of accesses at one or several network nodes, and
- Data security.

1.3.2 Hybrid Solutions

Today, ideal solutions are still impossible for two reasons:

1. The existing CAx systems have to be completely reorganized first in order to get open systems (for data exchange and for connections to DBMS).
2. There is a lack of a) Non-Standard and b) Distributed DBMS qualified for the engineering parts of CIM;
   a) Currently, pure RDBMS cannot attain the desired performance (see Fig. 1.7). Science and research have good approaches and prototypes of Non-Standard DBMS, but not yet proper systems for industrial use.
   In order to get a practical solution, hybrid solutions are used. Here, the engineering data models themselves are managed by a RDBMS as closed (unstructured) models, but their administration (e.g., version control) is handled explicitly by a SQL-based RDBMS. Figure 1.6 shows such a CAE archive prototype. This is a good compromise and also a good solution, at least as long as engineering data models can be unterstood by their own generator system.
   b) The unsolved trade-off between functionality, flexibility, and quality on the one hand, and performance on the other hand (i.e., lack of

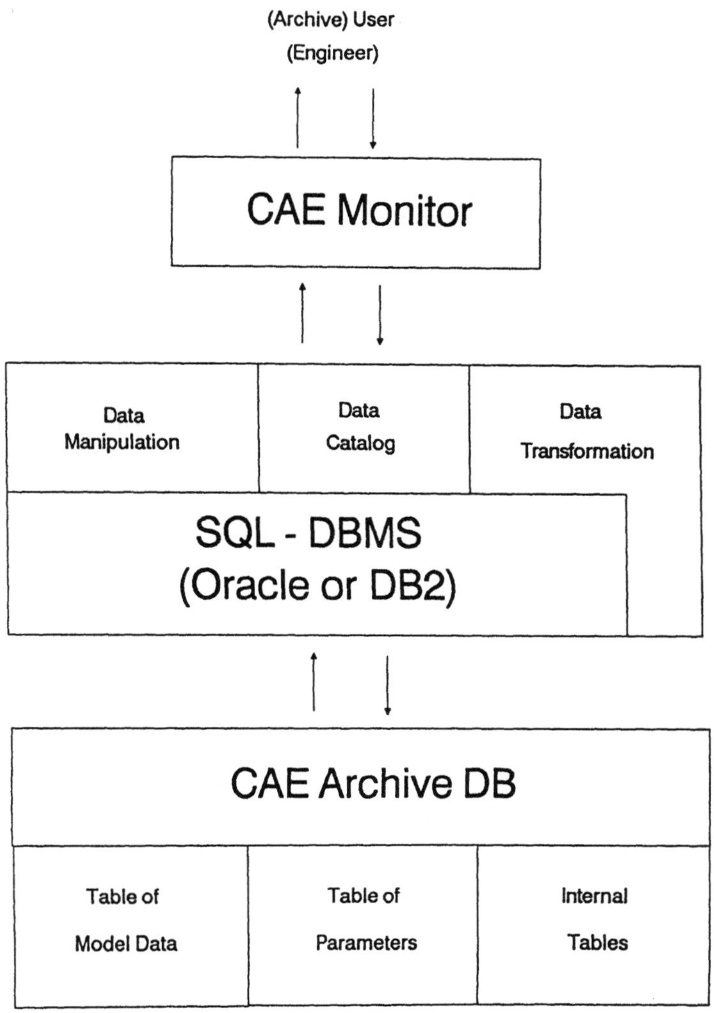

Fig. 1.6. Stucture of CAE archive

distributed DBMS), leads also to a similar compromise: the currently unavailable distributed DBMS is substituted by a so-called server-client solution, since the database problem can also be divided into two parts, namely into local incore modeling for rapid performance within an application system (client), and into central outcore modeling; that is, a central background database in which the entire information can be kept and managed at least with logical coherence (server).

Such a local incore modeler, RMI (Root Modeling Interface), enables the definition and manipulation of geometric data structures that are linked like networks. It has been designed as a model management system with high performance, a practically unlimited model storage capacity, and an integrated protection method.

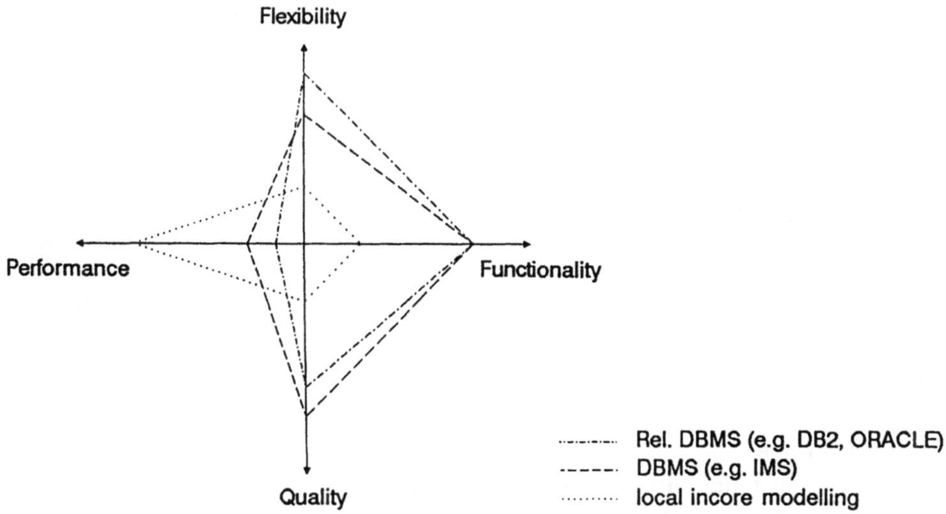

Fig. 1.7. Criteria of Database Management Systems

These local incore modelers show the way to the local subsystems of a future distributed DBMS, which will cover all four criteria of Fig. 1.7. Milestones on the way to a future Non-Standard DDBMS are e.g., XSQL, SQL++, NF^2, DAMOKLES, R^2D^2.

1.4 Organizational Procedures

1.4.1 Information Engineering

The state of the art shows that integrated global solutions are only possible via, or with the aid of, integrated database solutions. In other words, there must be an exact and complete division between the application and the database management system, and satisfactory and efficient coordination. The procedure must be correct to achieve this.

The new catchphrase "information engineering" is intended to define the new methodical procedure which manages the transition from data processing to information processing and retrieval. The key aspect in this change is the transition from being function-oriented to data-oriented. The information, i.e., data as their corresponding image on the carrier systems, are intrinsic, absolute resources. The objects, i.e., the data are now the centre of information systems and their structures are more stable than the functions to be used on them. This change is also to be found, for example, in the trend towards object-oriented languages and systems.

A comprehensive analysis of the flow of information and operations of a company is an indispensable prerequisite for a conceptional data model. Key words here are, for example, on-line system, time-sharing system, single user system, updating, etc.

A requirements analysis is the basis for the subsequent design of a data architecture which portrays the integration environment for the modular integration of all applications-related views of data into comprehensive data model that encompasses the entire company.

This model must be managed dynamically because updating (the possibility to modify and extend) the data architecture is indispensable for the practical value of this method. A rigid data architecture, on the other hand, would very quickly lose its importance as the binding orientation guideline. All function-oriented systems suffer to a greater or lesser degree from this aging problem (e.g., higher software maintenance effort).

The data are:

- entered once,
- kept free of redundancy (at least in the logical sense),
- edited as required, and
- presented at the correct time

on the basis of the company's integrated data model.

Another important effect of the data-oriented procedure is its positive influence on the development of application systems, because the data structure-oriented analysis of the applications organization provides a far more precise and consistent description of certain relations/associations than the function-oriented approach. The exact definition of the application system is more successful when using the data structure analysis than with purely function-oriented methods where data are only considered and analyzed implicitly in relation to functions and not with respect to their own logic in the data structure.

Furthermore, data modeling also makes an important contribution towards reducing communication problems between specialist departments and systems development - this is a major advantage (integration is teamwork within the company).

The data models are termed "conceptional" because they provide information about the significance and contents of the data and not about their technical realization. Definition of the data by abstraction not only supports their intrinsic stability but also their portability.

1.4.2 Management Aspects

All the technical features are dependent on the specialist and organizational competence of the company. Without the necessary qualified employees (including the top management) and without suitable structures in the organization and operational sequences and layout organization, it is not possible to realize effective CIM.

The basic requirement for successful CIM realization is an organization structure within the company that is tailored to the subject, and it is particularly important that there is direct contact with the top management. This is the only way to make autonomous and insular solutions a thing of the past.

It is also very important to choose the right time for the introduction and/or extension of new systems or concepts. Here it is right to say, the earlier the better; however, that is assuming there is a well-founded introduction strategy.

Further aspects to be given consideration here are the acceptance, motivation and training of the staff. These are continual and complicated processes which, generally speaking, cannot be initiated early enough (even if the systems are not yet ideal).

Advances in CIM will of course also mean the workplaces will become more demanding, and thus more will be expected of the staff members. At the same time, this is a great challenge to our training centres. The supply of a large number and, more importantly, well-qualified trainees, despite the animosity towards technical advances that is still to be felt on the labor market is, in our opinion, the key function towards achieving the planned CIM future.

A distinction must be made here between experienced CIM users and the people working on CIM realization (analysts, integrators) who, without doubt, provide a high percentage of the theory and computer experience.

The suppliers (universities, institutes, and industry), and also the users have been urged to work on this CIM future. The main task for the users together with their staff members is to adapt themselves to changes both in the operational sequences, and in their jobs.

This is a big issue and a challenge for everyone involved. The most important aspect ends this chapter:

integration is teamwork.

1.5 Further Reading

Eckert H., Längle D., Geissendörfer K., Dülfer R.: *RMI – A Geometry Data Management System for CAD/CAM.* Proc. CAD/CAM - Conference, Xian, 1987

Encarnacao J.L., Schlechtendahl E.G.: *Computer Aided Design, Fundamentals and System Architectures.* Springer, Berlin Heidelberg New York, 1983

W. Fischer: *A survey of ICES in CAE.* Proc. International ICES User Group Conference, 1987

Fischer W.: *Interactive Graphic NC Part Programming System.* Preprints 10th IFAC-World Congress an Automatic Control, 1987. Vol. 4, VDI-Gesellschaft.

Helmbrecht U., Böhnke U.: *Verwaltungsorganisation im CAE-Bereich mit Hilfe technischer Datenbanken.* Informatik-Fachberichte Nr. 171, Springer, Berlin Heidelberg New York, 1988

Kochan D.: *CAM, Developments in Computer Integrated Manufacturing.* Springer, Berlin Heidelberg New York, 1978

Lockemann P.C., Mayr H.C.: *Informationssysteme.* Springer, Berlin Heidelberg New York, 1978

Lockemann P.C., Schmidt J.W. (Hrsg.): *Datenbank-Handbuch.* Springer, Berlin Heidelberg New York, 1987

2 Database Technology

2.1 Concepts and Terminology

2.1.1 Characteristics of Database Systems

If there was one single cause that precipitated the development of database systems, it was the large-scale introduction of cheap direct-access mass storage in the form of magnetic disks. Up to then all mass-storage handling was in the form of sequential file processing. With the advent of disks, index-sequential files appeared at first, to be followed within a few years' time – at the end of the 1960s – by the first database systems. Since then, database technology has gone through a period of steady evolution towards mature software products as we know them today. And more importantly: the technology is well-prepared not only to survive all the new trends in hardware and system software such as novel processor architectures, workstations, computer networking, high-resolution screens, windowing and high-density magnetic and optic disks, but also to make the best possible use of these trends. The "best possible use" will, of course, be dictated as much by an application as by the underlying technologies. It is the first aspect, the usefulness to the application world, and in particular to engineering applications, which is the subject of Chap. 2.

No matter what the application is, a user has come to expect a certain quality from a database system. What are the characteristics that determine this quality? And what do we mean when we use the term "database system"? By way of discussing these questions, we shall establish the terminology that will be used throughout this book.

A *database system* (DBS) provides for the secure and reliable long-term management of data and for their retrieval according to individually (sometimes spontaneously) chosen criteria in a multi-user environment. From a structural point of view, a database system is a collection of stored data together with their description (the *database*) and a hardware/software system for their management, modification and retrieval (the *database management system* or DBMS). However, for the average user (as opposed to the database expert) this distinction between database system and database management system may sound artificial. Consequently, unless we address DBMS issues in the strict sense, we shall use the shorter term of database system.

There are eight key features that are commonly considered to make up the function and quality of a database system:

- data integration
- application-oriented data structuring
- consistency
- multi-user operation
- recovery
- transactions
- protection
- data independence

In contrast to classical file systems, where every file contains data whose structure matches exactly the requirements of one specific application program, database systems provide for the *integration of data* of all applications of an enterprise. Even though an application program may be restricted to exactly those data it needs and in the desired format, uncontrolled data redundancy causing storage overhead and update problems is avoided; only a minimum of data has to be replicated under system control.

The way in which data are structured when stored into and retrieved from a database reflects, from a user's (logical) point of view, the concepts and notations of his application, and is independent of physical storage considerations like blocks, tracks or bytes (*application-oriented data structuring*) .

As far as possible, a database system guarantees that the data it manages are as authentic an image as possible of (are *consistent* with) the application world. In other words, a database models the so-called *miniworld*, those facts that are considered to be of interest for a given application.

Due to the integrated management of data for many applications, database systems are usually expected to support *multi-user operation*. In this case, every user expects to have the impression of currently being the only one on the system. This requires access to data to be free from interference from *concurrent* users, or to be properly *synchronized*.

Large quantities of operational data represent valuable assets for the enterprise. Therefore, such data must be guaranteed to be *persistent*, meaning that a database system should provide *recovery* mechanisms that prevent the loss of information in cases of hardware and software malfunction, or when user operations are aborted.

As simple database operations like updating one single data item do not usually reflect meaningful application semantics, a concept is offered to define comprehensive units of work (called *transactions*). The database system then maintains certain properties with respect to consistency, concurrency and persistency for the entire sequence of operations that make up the transaction.

Protection mechanisms make sure that only properly authorized accesses to the database can be executed; they also include a component to define the authorization rules desired for a specific application environment.

Finally, database systems are constructed to provide *data independence*. With this feature, application programs are immune to changes in the way data are physically stored and to the strategies by which they are accessed.

For example, programs need not be changed or recompiled when new storage media are added to the system configuration, new access techniques included, or when additional applications require changes in the database structure.

2.1.2 Operation of Database Systems

Particularly because of data integration, applications cannot afford any longer (as with conventional file systems) to hide the miniworld semantics in the application programs. Instead, the semantics must be explicitly represented in the database itself. There are three distinct sources that together determine the database semantics:

a) data items as supplied by the users or their application programs,
b) classifications of and structures within the set of data items, and
c) additional explicit rules governing the data items and their structures.

These sources enter in various ways into the application semantics. The principal means for expressing them are the following.

- Every database system implements a so-called *data model* defining which *types* data items may have (i.e., which *value sets* for data items the system recognizes, and how these data items may be used), which *structure types* may be formed out of specific data items, which *rules* (restrictions) are observed for structures (because the database system observes them automatically, they are often referred to as data model inherent consistency), and the *operators* that may be applied to the database to insert, modify, delete and retrieve data items. These operators are *generic* in the sense that they are defined universally for all or certain classes of data, irrespective of their type. When applied, they have to be parameterized with the desired individual data items and structures whose types are known by then, so that the rules of the data model can be enforced.
  The set of operators forms the basis of the *data manipulation language (DML)* of a database system. There are several options as to how the DML is presented to the user. It may be embedded in one or more *host languages* (conventional programming languages like COBOL, PL/I, FORTRAN, PASCAL etc.). Alternatively, a stand-alone DML may be defined that incorporates the database operators and a set of other language features like arithmetic, loops etc. Stand-alone DMLs are mostly used for interactive database operation; they may also include graphical support.
- After a suitable data model has been chosen, and before DML operations may be performed, a *database schema* has to be defined for the database. The schema determines which data structure types and data item types out of those allowed by the data model will be used for this specific database. For example, a tabular data model will allow arbitrary tables as structures, with columns whose values are members of some predefined value sets. A schema for this data model will then define which specific tables may be used to insert and later on manipulate rows. A table is determined by

giving it a name, deciding on the number of columns and their names, and specifying the value set allowed for every column.
Operators for the definition of schemas are combined into a *data definition language (DDL)* that resembles the declaration facilities of high-level programming languages. Although some database systems do not require that the entire schema be defined before any data manipulation takes place, DML operations are only possible for structures that have been properly declared. Consequently, DDLs may be used stand-alone or, in some of the modern systems, interactively.

- Data model inherent consistency may not always be sufficient to reflect all the restrictions governing the miniworld. Hence, some database systems allow the definition of explicit rules for database structures and operations, mostly as predicates over data items or structures (so-called *consistency constraints* or *semantic integrity constraints*).
- Finally, after setting up the schema, data items may be inserted into the database and later on retrieved, modified and deleted, using the DML.

At first glance, one general schema for supporting all applications within an enterprise that use a database may seem sufficient. There is, however, a second side to the coin of data integration: A single application, or at least a group of applications, usually needs only part of the structures and data items defined in the schema, and may even want to see those organized differently from what the schema dictates (e.g., with some columns of a table omitted and the remaining ones reshuffled). To this end, database systems provide the concept of *subschema* or *view*. There are specific means in the DDL to define various subschemas for one schema, and in fact, applications often refer to a subschema and not necessarily to the schema itself.

Of course, although data are presented in a highly application-oriented form at the user interface, they finally have to be organized into bytes, blocks, tracks etc. at a lower level. While these aspects are purposely hidden from the user as far as possible, the database system has to provide the necessary transformation. We can thus distinguish three levels of data representation (cf. Fig. 2.1):

- the *external level*, represented by the set of subschemas (or *external schemas*) as used by end-users or their application programs, describes the database from the point of view of applications, one or a small number per subschema;
- the *conceptual level*, represented by the (one) general schema (or *conceptual schema*), describes the entire database in a uniform way; and
- the *internal level* describes the database in a way oriented towards its physical storage in secondary memory (a description internal to the database system).

As one may at least think of the internal level as being represented by an (internal) schema as well, this architecture has been termed the *3-schema-approach* to database systems.

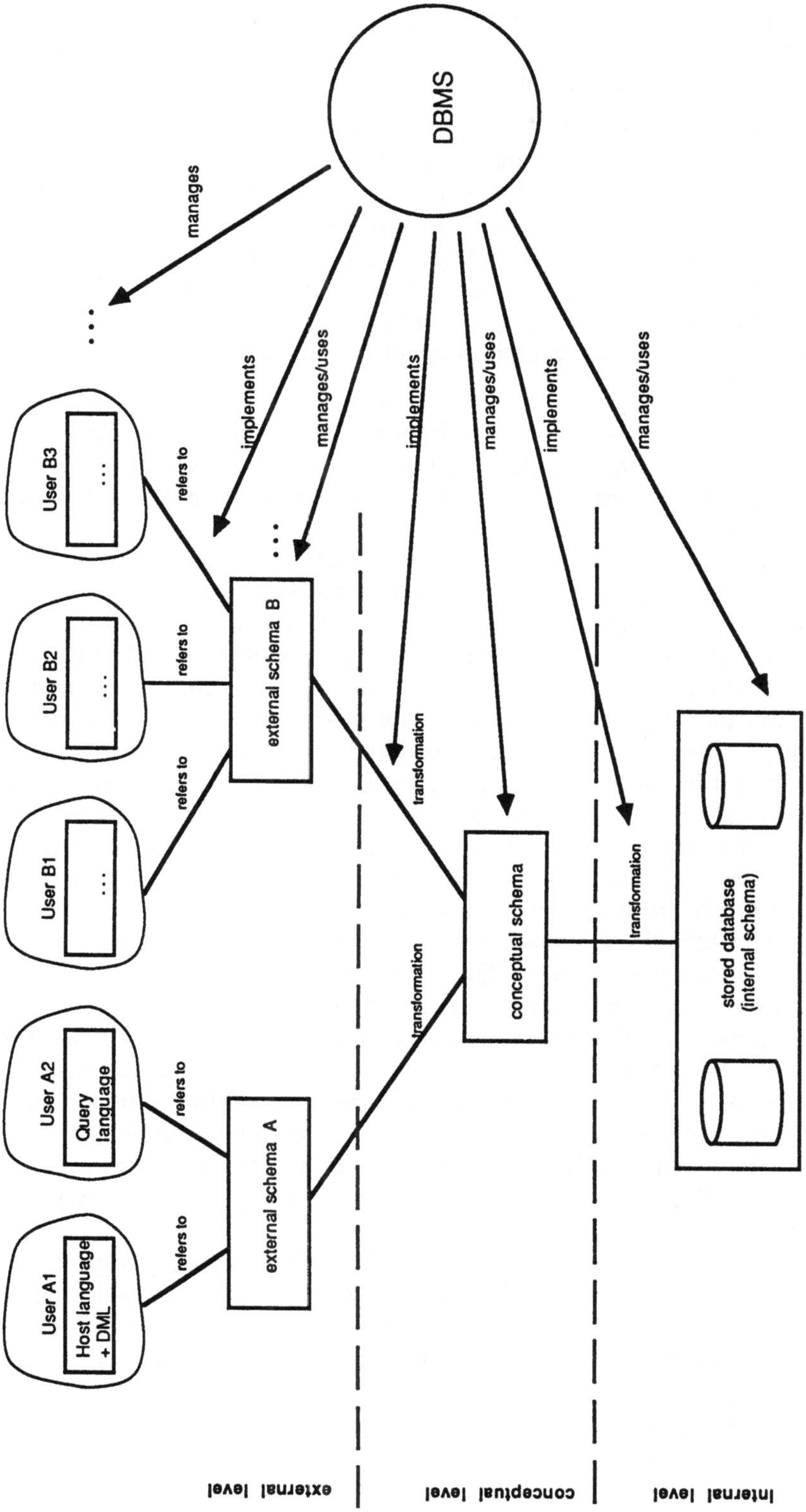

Fig. 2.1. Three-schema-approach to database systems

The requirement to store not only the user data, but also their description, and to translate from external via conceptual to internal representation, results in a *stored database* that contains various classes of data:

- *user data* as supplied via DML operations,
- schema data (data descriptions, *metadata*) as supplied via DDL operations,
- *system data* that are created and maintained by the database management system to handle all necessary transformations,
- administrative data like consistency constraints or authorization rules, and
- *access paths* that are constructed and maintained by the DBMS to perform DML operations efficiently (e.g., hash tables and indexes to retrieve data according to particular criteria without having to scan the whole database sequentially).

Whereas the user data comprise, from a user's point of view, the only data useful to him (they are, therefore, called the *primary data*), the other classes (the *secondary* – or auxiliary – *data)* often make up a significant portion of the entire database.

While database systems allow the expression of application semantics in a problem-oriented way, the task of defining a database schema and the accompanying constraints (called *database design*) remains a difficult process. Various methodologies and tools have been proposed, especially for the business and administrative types of application, but casting the miniworld information into the framework of a data model and consistency constraints will always require intuition and experience. We will not try to review the state of the art in database design, but just give a brief explanation of how the three levels of a database system are affected.

As for the upper two levels, every application has to determine and formulate the requirements for its subschema. All these views then have to be combined to obtain the general database schema in terms of the DDL (*view integration*). Once the schema has been defined, all necessary subschemas are derived by using appropriate DDL operators, and will hopefully match the initial requirements. As there may be trade-offs and even contradictions among different requirements, schema and subschema design is usually an iterative process involving both application and database specialists. In larger environments, a so-called *database administrator* (DBA) coordinates and mainly executes the database design, and later on controls all database activities (note that database administrator is a role rather than a specific person: depending on the evironment, it may be filled on a part-time basis or by an entire department).

To deal with the third level, database administration is responsible for *physical database design* (in contrast to *logical* design as explained above). This task involves the setting of all system parameters that influence the internal data representation, access paths, and level-to-level transformations. When the system performance becomes unacceptable, some of these parameters may have to be redefined, sometimes causing a database *reorganization*.

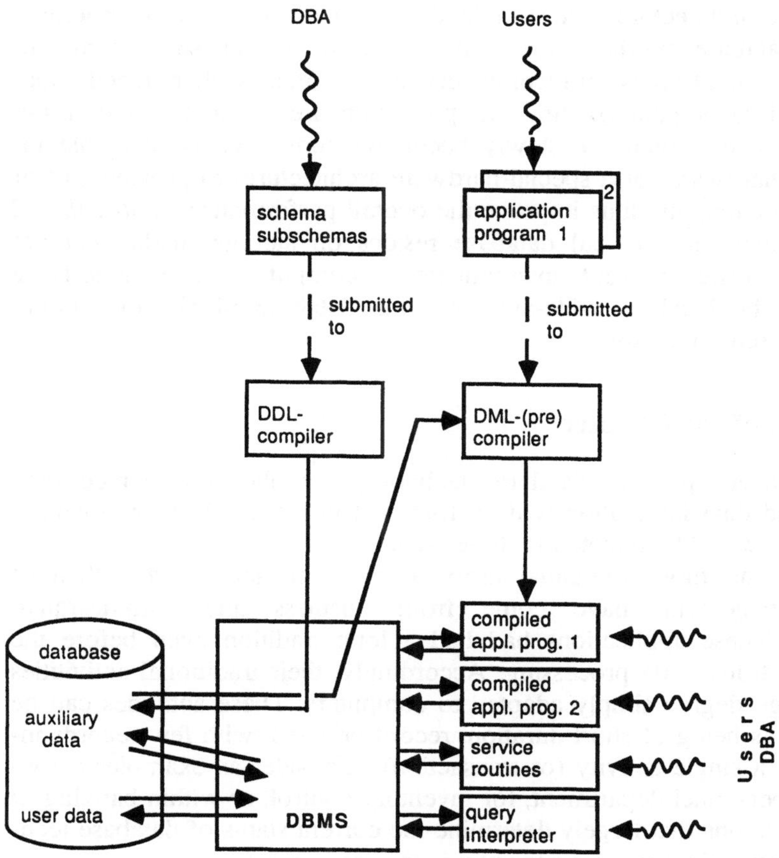

Fig. 2.2. DBS installation and operation

Figure 2.2 summarizes the main aspects of database system installation and operation:

- Schema and subschema are defined using some DDL, and are compiled into some system-internal format which is stored in the database.
- Host-language application programs including DML statements are compiled, either in one step (combined host language/DDL compiler) or, more commonly, in two steps (DML precompiler/host language compiler). Compiling DML statements requires references to the schema information stored in the database.
- Users run compiled application programs which read from and write to the database via the DBMS.
- Ad hoc queries and commands are handled by an interpretive query facility supplied together with the DBMS. From the DBMS point of view, the interpreter looks like any other application program.
- In the same way, a variety of *service routines* for database administration, report preparation etc. are provided.

Finally, two architectural issues in database systems are worth mentioning. Traditional database products are constructed to run in single-computer environments (mainframes, minicomputers, and – often with reduced functionality – microcomputers); they are pure software products. In contrast, *database machines* which are slowly becoming commercially available include special hardware (or a special hardware architecture) to provide part of the DBMS function, and thus improve the overall performance. In *distributed database systems*, one logical database resides on storage media that are controlled by different (but interconnected) computers. The underlying network may be local or wide-area, but the database distribution should remain transparent to the user.

2.1.3 Outline of the Chapter

For the engineer applying database technology, application-oriented data structuring and data integration will be foremost in his mind. Consequently, much of Chap. 2 will be devoted to these issues.

The early and most influential users of file management, and then of database management, have come from business and administrative applications. These applications had had a long tradition even before the advent of electronic data processing. Accordingly, their traditional activitities were to a large degree simply adapted to computers. These activities can be characterized as being of short duration, record-oriented, with few records involved within a single activity (or transaction). Consider as examples activities within a personnel department, for inventory control, or within banking. It is these applications that largely determine the current status of database technology, and hence the products currently in the marketplace.

Very early it was found that records could not always be viewed in isolation or as members of just a single set of objects, but that they entered into more complex interrelationships among themselves. At first these were considered to be hierarchical in nature as, e.g., in a business organization or a bill of parts. The result was what is called the hierarchical data model which, because of its limitations, we shall cover only very briefly.

Later on, when data communication made it possible to maintain and access all the data of an enterprise – perhaps geographically distributed – within a single repository, hierarchical data organization became something of a liability unless one tolerated a high degree of redundancy. Consider a personnel record in an enterprise that may be part of a file kept by the personnel department, another file kept by the department where the respective person works, a third file for a project, and at times a fourth file of all persons on leave. Or to give another example, take an article record that may variously appear in an inventory file, production file, order file or sales file. If integration is supposed to mean that such a record should be accessible via any of the files, and that any change to the record should instantaneously be visible in all files, a tree-like hierarchical organization will not suffice. What is needed instead is a network-like data organization. We shall treat the corresponding data model in considerable detail.

Record-at-a-time processing of files may sound fine in a batch-processing environment. In a highly interactive environment where users query the database from a terminal on a spontaneous basis, collecting a number of records with some common properties from the database and displaying them together on the screen may be more desirable. This idea gave rise to a different kind of data organisation, one that collects the records into tables or – as they are called – relations. Relations are akin to files. However, records are selected on the basis of their contents, files can selectively be merged on the basis of contents, while redundancies may still be avoided by proper design. The corresponding relational data model is another subject of this chapter.

As we shall see, representing engineering situations in these data models is far from being natural and efficient. What is needed are data models that permit the manipulation of collections of records that are somehow inter-related but otherwise dissimilar. Underlying all modern trends to data models for database systems is the notion of "object". This notion has gained currency in fields as diverse as engineering applications, office automation, and knowledge representation. Accordingly, there are various interpretations of what an object is, and hence, a variety of approaches to data models. We shall discuss a few of the more promising ones.

Once the engineer has gained fluency with the data structuring capabilities of a database system he will start to appreciate all of their other characteristics – or demand them if they are not readily available. Not surprisingly, they do not come for free, but exact a certain price in terms of performance. Consequently, the engineer should be familiar with the aspects – though not necessarily with the technical solutions – of consistency, concurrency, recovery, transaction management, protection, and data independence. We shall introduce these issues and discuss them mainly from a user's perspective.

2.1.4 A Running Example

To familiarize the uninitiated reader with as complex a technology as database systems is by no means trivial. To ease the task we shall accompany the discussion of the various issues with the same running example in order to illustrate their main features, strengths, and weaknesses. The example is taken from VLSI design. Consider Fig. 2.3 where a cell (CELL3) has been designed using two earlier defined cells CELL1 and CELL2. CELL1 has four external ports such that several of its instances may be placed directly side by side. The topology of CELL3 is defined by packing four CELL1 instances (numbered I1 through I4) in one row and placing a single instance of CELL2 (named I5) above the row, interconnecting it with three of the CELL1 instances by paths P2, P3 and P4. One external port of the CELL2 instance and one each of the marginal CELL1 instances are extended (paths P1, P5 and P6) to the boundary of CELL3 to form its external ports. CELL1 and CELL2 have their own topologies of which we only show the conductor paths. We also assume that, as part of their topologies, all their ports at the left-hand and upper edges are inputs, all others outputs.

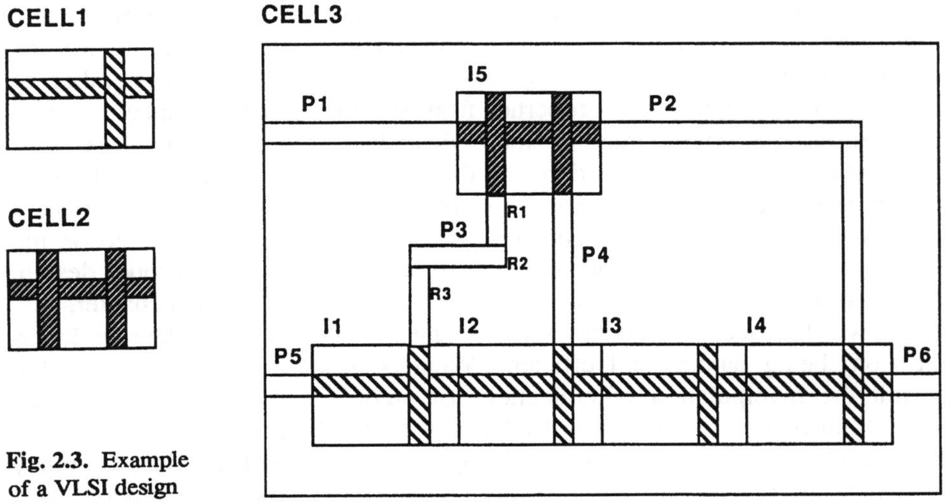

Fig. 2.3. Example of a VLSI design

There is also a certain geometry to each of the cells, such as the sizes and the positional coordinates of the CELL1 and CELL2 instances, the size of CELL3, and the sizes, positions and forms of the conductor paths. In fact, the paths may be considered elements in their own right, to be described by their composition from rectangles. Again, rectangles may be defined as elementary geometric symbols, though without the rich semantics of a cell or a path, and parameterized by length, width, position of origin, and orientation.

To the topological and geometrical information one should add technical and administrative information. The former will, in our example, include for cells their technology (e.g., CMOS, nMOS, ECL), function, number of inputs and outputs; for paths their resistance and capacitance. The administrative information consists of the name of the object, the date of completion, and a release indicator.

The designer, in constructing the cells and paths, must observe a few rules.

1. The outputs of a cell may only be connected to inputs of other cells or to a single output of an enveloping cell.
2. The input of a cell may only be connected to a single output of one other cell or to a single input of the enveloping cell.
3. An output of a cell may only be connected to f (where f is the fan-out factor) inputs.
4. The width of a path must range between *wmin* and *wmax*, and the length must not exceed *lmax*.
5. No two paths may intersect.
6. All cells in a design use the same technology.
7. No two cells may overlap.
8. A cell cannot contain a cell of its own type.

The attentive reader may have noticed that cells are designed as prototypes, instances of which may readily be duplicated and arbitrarily positioned.

Topological, technical and administrative properties can be associated with the prototypes and need not be repeated for the instances. The same is true for the geometric property of size, whereas the other geometric information must be assigned to each instance. Paths do not appear to be prototypical, whereas rectangles again are.

2.2 Data Organization and Manipulation: Current Status

2.2.1 The Classical Data Models

2.2.1.1 The Hierarchical Data Model

The basic concepts of the hierarchical model are as follows. A **database** consists of an arbitrary number of similarly structured hierarchies (called **database records**) of records in the traditional sense (that here are called **segments**). The hierarchical structure itself may be pictured as a tree with a superimposed class structure. The root of a tree is called the **root segment** of the tree, and its immediate successors its **children**. The set of children is itself partitioned into one or more classes, each with identically structured segments, i.e. segments of identical **segment type**. The segment structure is defined by a set of **fields**. Segments within a class are referred to as **siblings** and may (but need not be) individually identified by a **key**. The root itself is the **parent** of its children.

Such a one-level hierarchy may be extended to arbitrary depth by considering each of the children as the root of a subtree with the same structure as described before, and by continuing to do so level by level. The root segments of database records are themselves of identical type and, hence, also form a class with the characteristics discussed above.

In order to describe the structure of a database it suffices to describe the structure of a database record. That structure can be formulated as a hierarchy of segment classes, with each class represented by its respective segment type. Accordingly, the database schema is a tree of segment types. We follow the usual conventions by representing the schema in graphical form (Fig. 2.4). There is but one restriction (rule) to be observed: children of a segment must not be of the same type as their parent.

A database that satisfies such a schema is formed as follows. Take the root segment type and replace it by a set of segments of that type. For each segment in this set continue the process just described for each of its child segment types. Observe that empty sets are quite legitimate. Figure 2.6 gives an example.

With all this having been said we should be in a position to formulate our example of Sect. 2.1.4 as a hierarchy. This should pose no problems: After all, is defining cells as a conglomerate of simpler cells not a hierarchical process?

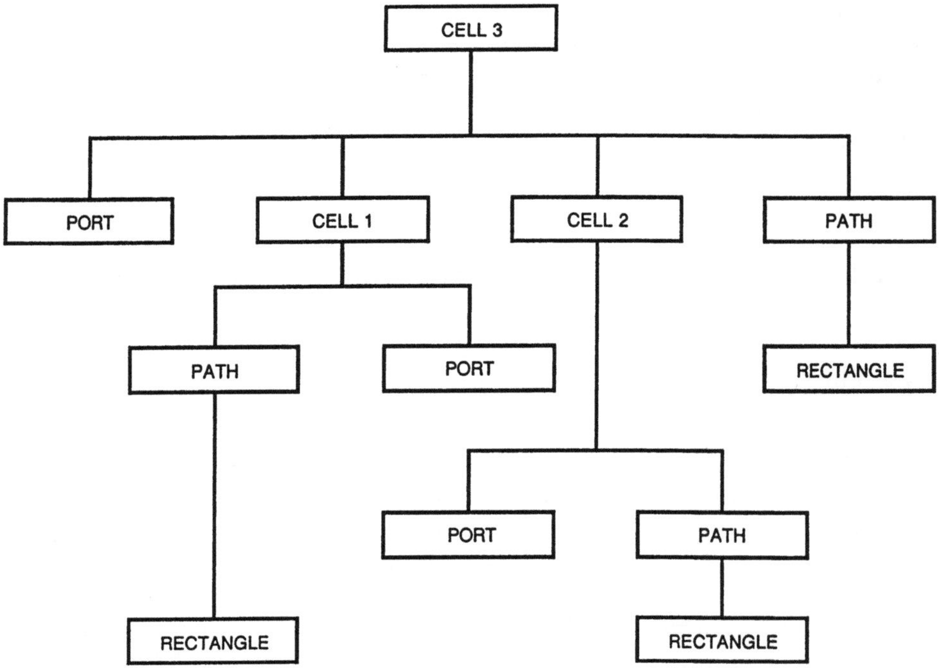

Fig. 2.4. Hierarchical schema for the design in Fig. 2.3

CELL 1	(identifier: string; completion: date; released: boolean; technology: (CMOS, nMOS, ECL); function: string; no_of_inputs: integer; no_of_outputs: integer; size: integer; x_position: integer; y_position: integer).
CELL 2	(identifier: string; completion: date; released: boolean; technology: (CMOS, nMOS, ECL); function: string; no_of_inputs: integer; no_of_outputs: integer; size: integer; x_position: integer; y_position: integer).
CELL 3	(identifier: string; completion: date; released: boolean; technology: (CMOS, nMOS, ECL); function: string; no_of_inputs: integer; no_of_outputs: integer; size: integer; x_position: integer; y_position: integer).
PATH	(identifier: string; resistance: real; capacitance: real).
RECTANGLE	(identifier: string; length: integer; width: (wmin..wmax); x_origin: integer; y_origin: integer; orientation: integer).
PORT	(identifier: string; inout: (I, O); x_position: integer; y_position: integer).

Fig. 2.5. Details of the segment types of Fig. 2.4

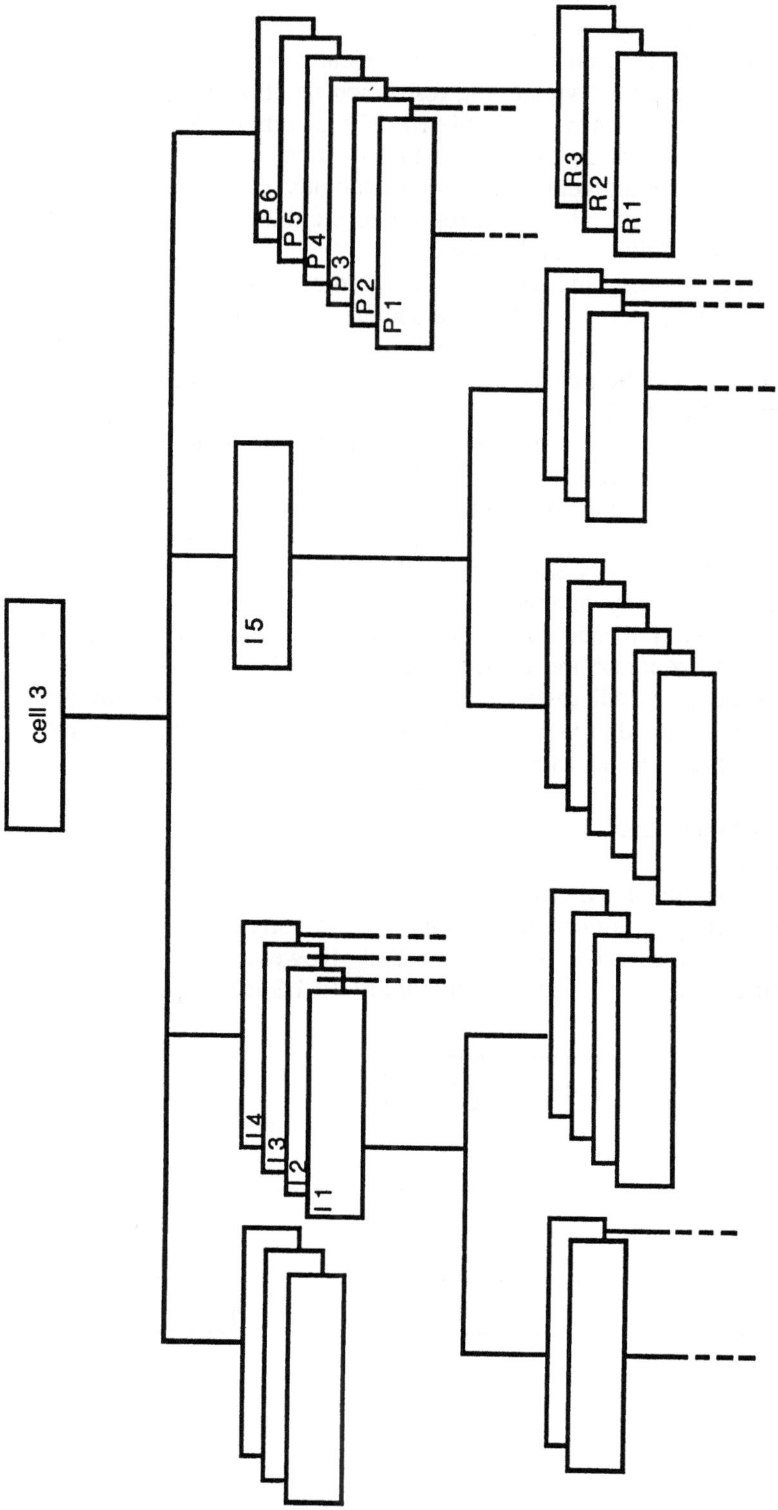

Fig. 2.6. Database for the example of Fig. 2.3 satisfying the schema of Fig. 2.4. Field values are not shown, and neither are the rectangle segments except for Path P3. Notice that for all instances I1 through I5 the subtrees are identical

What are the segment types? Are they cells and paths? Even though this may sound like the natural approach, the hierarchical model blocks it because cells as children of cells would violate the restriction mentioned above. Consequently, we shall have to distinguish between cells of types CELL1, CELL2, and CELL3. No such problem exists with paths, since – contrary to cells – these are not defined recursively but from rectangles. Since ports play an important role in the design rules, we propose to include them as a separate segment type. Figure 2.4 illustrates the result.

As the next step in designing the schema we must decide on the precise segment structures, i.e., which fields to associate with each segment type. These are derived from the list of geometrical, technical, and administrative properties in Sect. 2.1.4 (whereas the topological properties are reflected in the database structure). Figure 2.5 gives the details of the segment types in the form: segment type name followed by the field names and their value sets (in parentheses). Finally, Fig. 2.6 is a graphical illustration of the database for Fig. 2.4 that has been constructed according to the rules given above.

The reader will doubtlessly agree that the schema is less than satisfying. First of all, as Fig. 2.6 clearly demonstrates, the database is highly redundant, since the subtrees and many of the field values are repeated for each of the four CELL1 instances I1 through I4. This is due to the fact that our schema does not allow us to separate the prototypical information on cells from the information specific to a cell instance. Redundancy, however, is something that should usually be avoided because – besides the waste of storage space – maintaining all copies in the same state at all times is an expensive and volatile affair.

With these observations in mind, it seems rewarding to attempt an alternative schema design that explicitly recognizes the prototypes. Figure 2.7 shows a conceivable result (CELL_INSTANCE refers to the cell instances from which the prototype is composed), Fig. 2.8 the corresponding database. Clearly, the redundancies have been eliminated. However, we paid a new price. The schema no longer reflects the fact that CELL3 is composed of cells

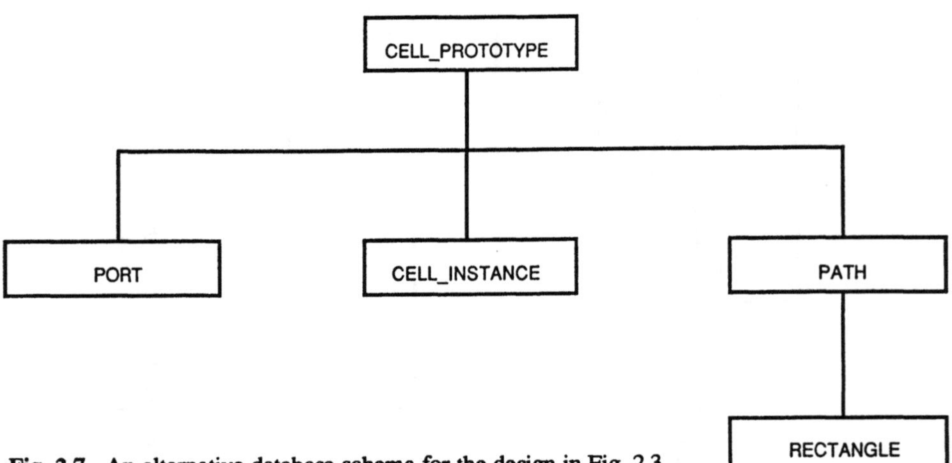

Fig. 2.7. An alternative database schema for the design in Fig. 2.3

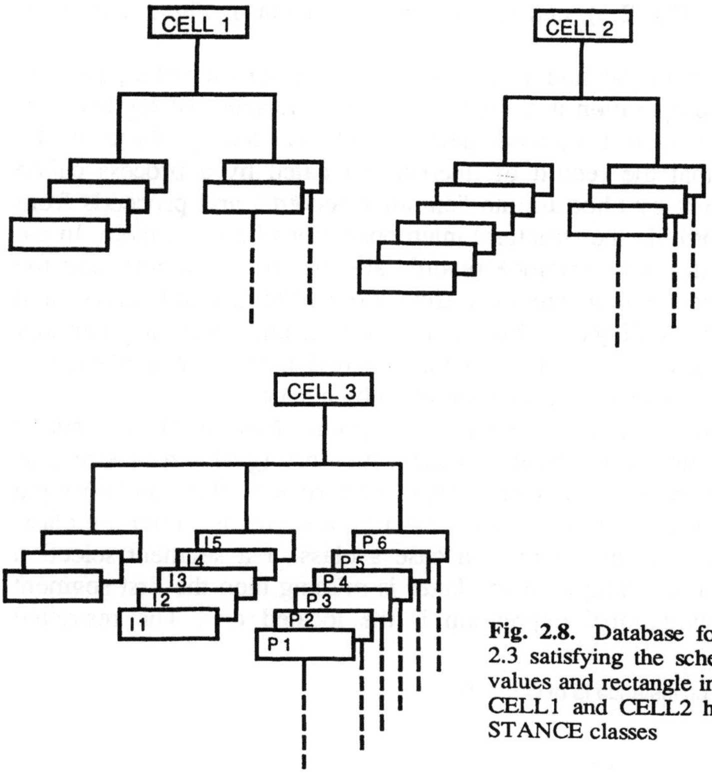

Fig. 2.8. Database for the example of Fig. 2.3 satisfying the schema of Fig. 2.7. Field values and rectangle instances are not shown. CELL1 and CELL2 have empty CELL_INSTANCE classes

of prototypes CELL1 and CELL2. If we wish to find this out we must recognize that cell instances I1 through I4 are of cell type CELL2, and accordingly search the database for the corresponding database records.

In other words, some of the topological information has not been preserved in the database structure and thus must be reconstructed procedurally. In fact, this is not the only topological information missing from either Fig. 2.4 or Fig. 2.7. Which paths are connected to which ports of which cells is not reflected in the database, nor is which of the cells adjoin one another without an intermittent path. Both could be reconstructed from the database only by "circumstantial evidence", namely by inspecting and comparing the (x,y) positions (or instead, one would have to add appropriate fields to the segments).

We conclude: The hierarchical data organization reflects at best the hierarchical design process but not the topological information – the latter is more network-like. Other information cannot be derived directly either, such as a list of instances for each cell prototype. Again, it must be collected by traversing the database records.

How well do the schemas enforce the design rules in Sect. 2.1.4? The DBMS guarantees that any update operation never results in a database that cannot be derived from the schema. In other words, which of the design rules are reflected in the schema? The answer is just that part of rule 4 concerned with path width (see Fig. 2.5), and rule 8 for the schema of Fig. 2.4 (but not Fig. 2.7!).

Data manipulation in the hierarchical model is **record-oriented**, i.e., the subject of a database operation is a single record (in the sense of segment). A characteristic of all record-oriented data models (including those to be discussed later) is that the record of interest is located by a process called **navigation**: One starts by choosing an "entrance record", and proceeds from there record by record via the structural interconnections between them. In the hierarchical model the only entrance records are the root segments, and the only available interconnections are those from a parent to its child classes, and within a class via the siblings, or from a parent to a child with a given key value. Once the desired segment has been reached it may be subjected to reading, modification, deletion, or augmentation by a successor.

The examples of Fig. 2.9 illustrate this point. Navigation is usually combined with reading the segment into main memory (*get*) and proceeds as follows. As one walks down the hierarchy, one alternates between choosing a class identified by its segment type) and selecting a segment within the class. There are also special interpretations in case a class or a segment selection have been omitted. For example, if the latter is missing then the first segment that satisfies the navigational expression is the located one. The restricted

"Give me all information on instance I5 of CELL 2"

```
get         CELL 3 (identifier = 'cell 3')
            CELL 2 (identifier = 'I5')
```

"List all ports of CELL 1"

```
get         CELL 3 (identifier = 'cell 3')
            CELL 1                          /* locates the first instance */
            PORT                            /* locates its first port */
while       class_not_exhausted do
getnext     PORT                            /* locate the next port */
```

"Delete path P2"

```
get         CELL 3 (identifier = 'cell 3')
            PATH (identifier = 'P2')
delete                                      /* deletes the
                                            entire P2-subtree */
```

"Insert a new cell CELL 2 "

```
insert      CELL 3 (identifier = 'cell 3')
            CELL 2                          /* adds the segment in work
                                            space to the class */
```

Fig. 2.9. Operations in a database satisfying the schema of Fig. 2.4

capability that forces the navigation to start with a root node is the main obstacle to efficiency in a hierarchical database if one frequently wishes to access segments all the way down a tree and close to the leaves.

Navigation also means that a database according to Fig. 2.4 does not directly reflect the concept of a cell; instead, the user must be aware that the concept is made up of several segment types, and the corresponding information must be collected by traversing the database records. The notion of object mentioned in Sect. 2.1.3 is precisely an attempt to overcome this weakness.

2.2.1.2 The Network Data Model

Two of the shortcomings of the hierarchical model – failure to account for a network-like topology, and a highly restricted set of entrance records – are overcome by the network data model. To start again with the structure concepts: A **database** consists of an arbitrary number of **records** and interrelationships between them. Each record is of a certain **record type** that determines its composition from fields named by **attributes**. The relationships are of a special kind: They establish a function from one record type, say R1, to another, say R2, where R2 is not equal to R1. This means that under a particular relationship only a single record instance of R2 may be associated with a record instance of R1, whereas an R2 instance may relate to any number of R1 instances.

The network data model considers the converse of this function and partitions it into subrelations such that each subrelation identifies a single R2 instance together with all its associated R1 instances. Such a subrelation is called a **set**, the R2 instance the **owner** of the set, and any R1 instance in the set a **member** of the set. Sets have a **set type** that is determined by owner record type and member record type, as well as a set name (more precisely: a set type name). Figure 2.10 illustrates the set concept for the relationship between a path and the rectangles that make it up (this clearly is the converse of a function). The members of a set may be ordered according to a variety of criteria.

From the definition of set it becomes clear that a particular record may appear at most once within a single set instance of a given set type, either as the owner or as a member (this clearly is a rule in the sense of Sect. 2.1.2). Consequently, a record uniquely identifies an instance of a given set type. Therefore, set instances need never be named. On the other hand, the same record may participate in any number of sets as long as they are of mutually different types; in some of these it may be a member, in others an owner. In particular:

- Suppose that record type R is member type in a set S1 and owner type in a set type S2. Then R establishes a two-level hierarchy. Figure 2.11 gives an example.
- Consider record type R to be an owner type in two set types S1 and S2 or, instead, a member type in both. Then R establishes a forking or merging

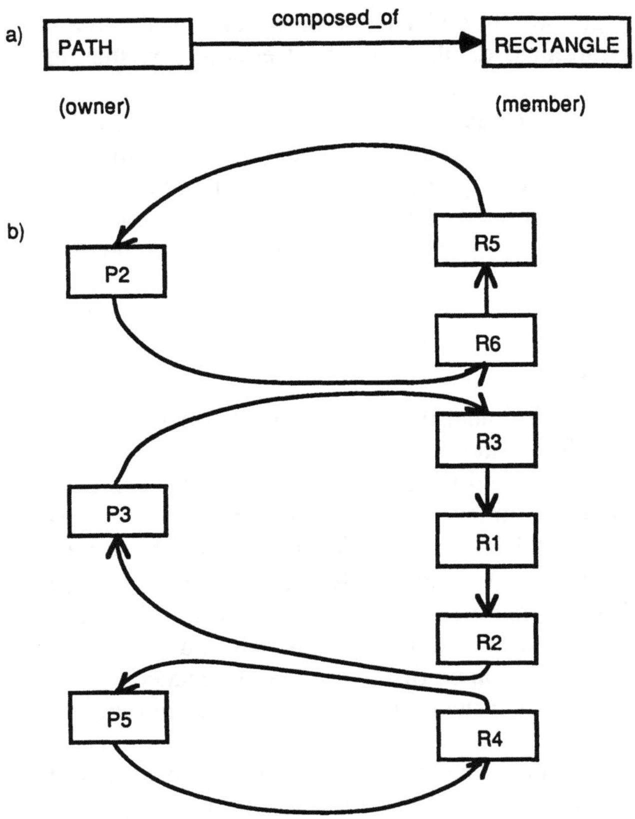

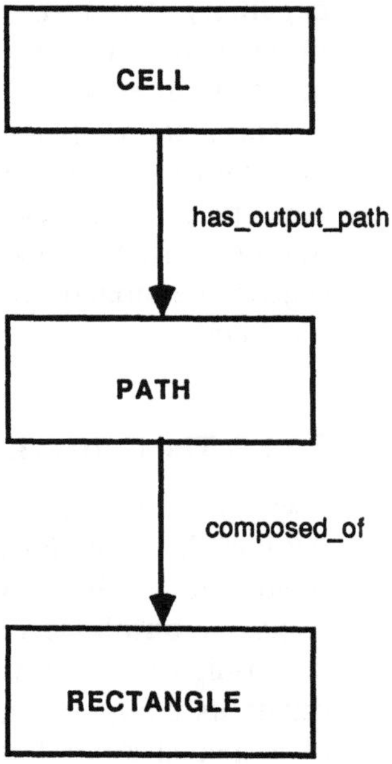

Fig. 2.11. Hierarchical arrangement of sets

Fig. 2.10a,b. Illustration of the set concept. **(a)** Schematic representation. The set type is symbolized by an arrow from the owner record type to the member record type; the arrow is marked with the set name. **(b)** A sample database for the set type. Sets are usually represented graphically by circular lists

point, respectively; both are the essential ingredients for building a network. Figure 2.12 illustrates this for a particularly simple case. Notice how redundancy (records P2, P3, and P4) is entirely avoided.

Although the set concept definitely adds flexibility as compared to the hierarchical model, it has its drawbacks. We summarize the rules:

1. owner and member record types must be different;
2. no attributes may be associated with a set;
3. a functional relationship (or more precisely, the converse) must underlie the set; and
4. no more than two record types may participate in a given relationship, i.e., the relationship is binary.

These rules may cause difficulties when a schema is to be designed. If they do, all these cases are solved in the same fashion: Consider the relationship that one wishes to model by a set to be an object in its own right, represent it

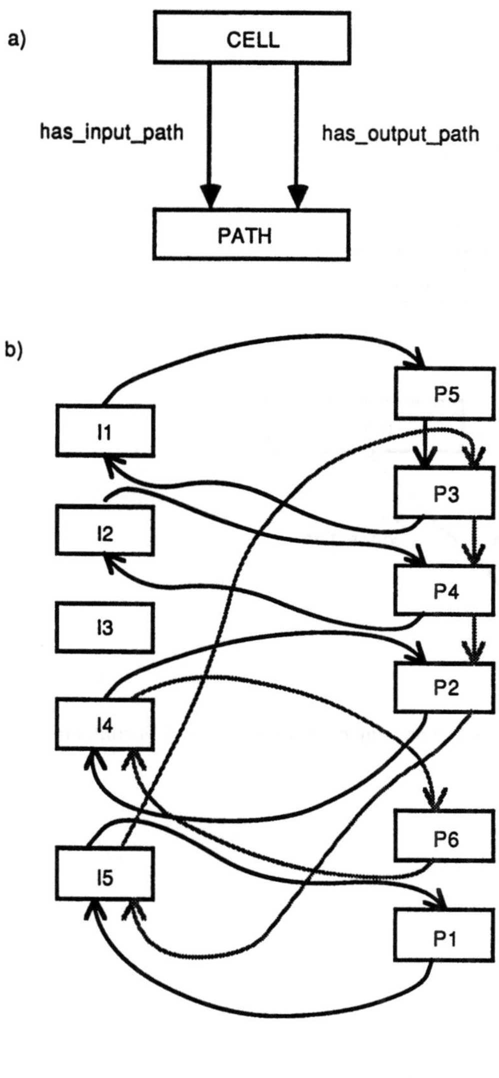

◀ **Fig. 2.12a,b.** Simple network arrangement of sets. (**a**) Schema. In general, merging need not occur immediately after forking. (**b**) Associated database. Solid lines: has_input_path. Grey lines: has_output_path

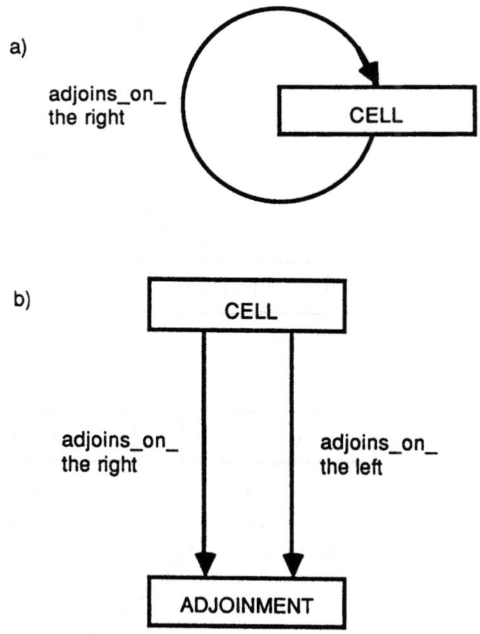

Fig. 2.13a,b. Introduction of a link record to separate owner and member record types. (**a**) Forbidden constellation. (**b**) Redesigned schema

by a separate record type (**link record** type), and establish set types from each record type participating in the relationship to the link record type. Figure 2.13 gives an example for rule 1. If the relationship in Fig. 2.13 had simply been "adjoins" then requirement 3 would also have been violated; the solution, however, remained the same. Figure 2.14 illustrates rule 4 and is a modification of Fig. 2.12 for one of the sets: We wish to represent the fact that a cell is connected to a particular (input/output) path via a particular (input/output) port.

One neat side effect of the set concept is that it allows partitioning of the set of all records in the database into subsets, provided the records of the subset are all of the same type. Particularly useful are subsets that encompass all records of a given type, such as all cells, e.g., in order to process all of them together. To do so, an artificial owner is needed. The network model

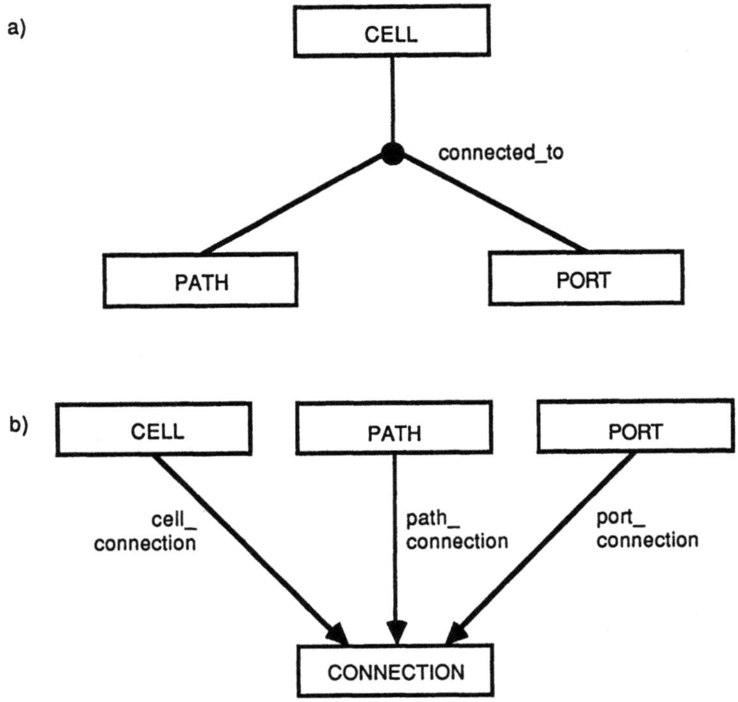

Fig. 2.14a,b. Introduction of link record to resolve a three-place relationship. (**a**) Forbidden constellation. (**b**) Redesigned schema

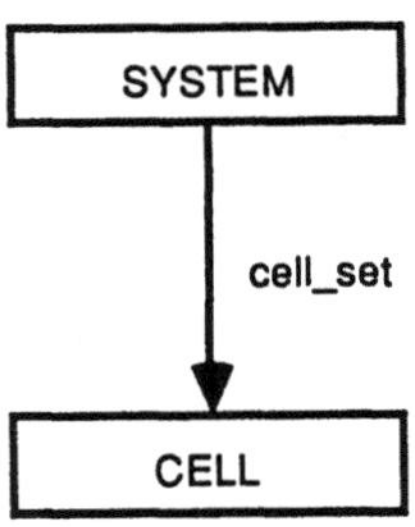

Fig. 2.15. Example of a system set schema

reserves a particular SYSTEM record as a stand-in owner that remains invisible to the user (Fig. 2.15). The corresponding sets are called **system sets.** Since there exists only a single SYSTEM record, only one set instance per system set type is possible (hence the term "singular set", which is also used).

A database that satisfies a network schema looks as follows. For each record type in the schema there may exist an arbitrary number (including zero) of records. A record may participate in any number of sets of the types with which its record type is associated in the schema. Each record automatically defines a set of each set type in which its record type is an owner record type, and becomes the owner of that set. For example, in Fig. 2.12 cells I1, I2, and I3 are also owners of sets of type has_output_path, although no member records exist for them, and similarly for I3 and has_input_path. One may even go further and require that a record always be a member of some set of a set type in which its record type is a member record type (often referred to as the MANDATORY AUTOMATIC option). Take Fig. 2.10 as an example. A rectangle makes sense in the database only if it is part of some path; hence, we declare RECTANGLE to be a MANDATORY AUTOMATIC MEMBER of composed_of (see also Sect. 2.2.2.3).

With all these introductory remarks and examples we should now be in a position to draft the schema for our VLSI example. Like in the hierarchical representation, we consider cells, paths, rectangles and ports as the information of primary interest to us, to be represented by record types. In order to distinguish the prototypical information about cells from information specific to cell instances (mainly positional data) we introduce separate record types CELL_PROTOTYPE and CELL_INSTANCE. The same argument could be applied to ports. However, we elect to use a single record type PORT that could be interpreted differently in different contexts. If it is considered in conjunction with CELL_PROTOTYPE it reflects the prototypical port information. On the other hand, if used in conjunction with CELL_INSTANCE it reflects the ports of this instance (the positional data of the port can be derived from those of the cell instance).

After having established the record types we introduce the set types. The network model leaves us rather wide latitude in doing so because we may declare one or more sets between any pair of record types. Which of these are really meaningful can be judged on a very straightforward basis: Try to assign a name to the set type that expresses its semantics very naturally; if none can be found the set type is probably superfluous. If any of the remaining set types violates the restrictions, we break it up into two types and a link record type. Finally, we decide which of the record types should be subjected to a system set.

In the present case we wish to identify separately the input ports and the output ports for each cell prototype. This gives rise to two set types from CELL_PROTOTYPE to PORT. There are also two relationships between CELL_PROTOTYPE and CELL_INSTANCE. One (has_instance) is a generic relationship used to list all instances of this cell type in the database. The other (contains_instance) is a part relationship and gives for a prototype (e.g., CELL3) all cells (here CELL1 and CELL2 instances) from which it has been constructed. We also wish to determine which port of which cell instance (i.e., which port instance) is connected to which path. In essence, this establishes a three-place relationship which must be resolved according to Fig. 2.14. Further, the database should describe the composition of paths from rectangles (Fig. 2.10) and the adjoinment of cell instances (resolution according to Fig. 2.13). Figure 2.16 gives the result of this analysis, Fig. 2.17 lists in detail the composition of the record types. Also, by comparing Fig. 2.16 to Fig. 2.5 we notice that some record attributes have been dropped (e.g., no_of_inputs, no_of_outputs in CELLs), because they may easily be computed from the sets. The link records contain no values.

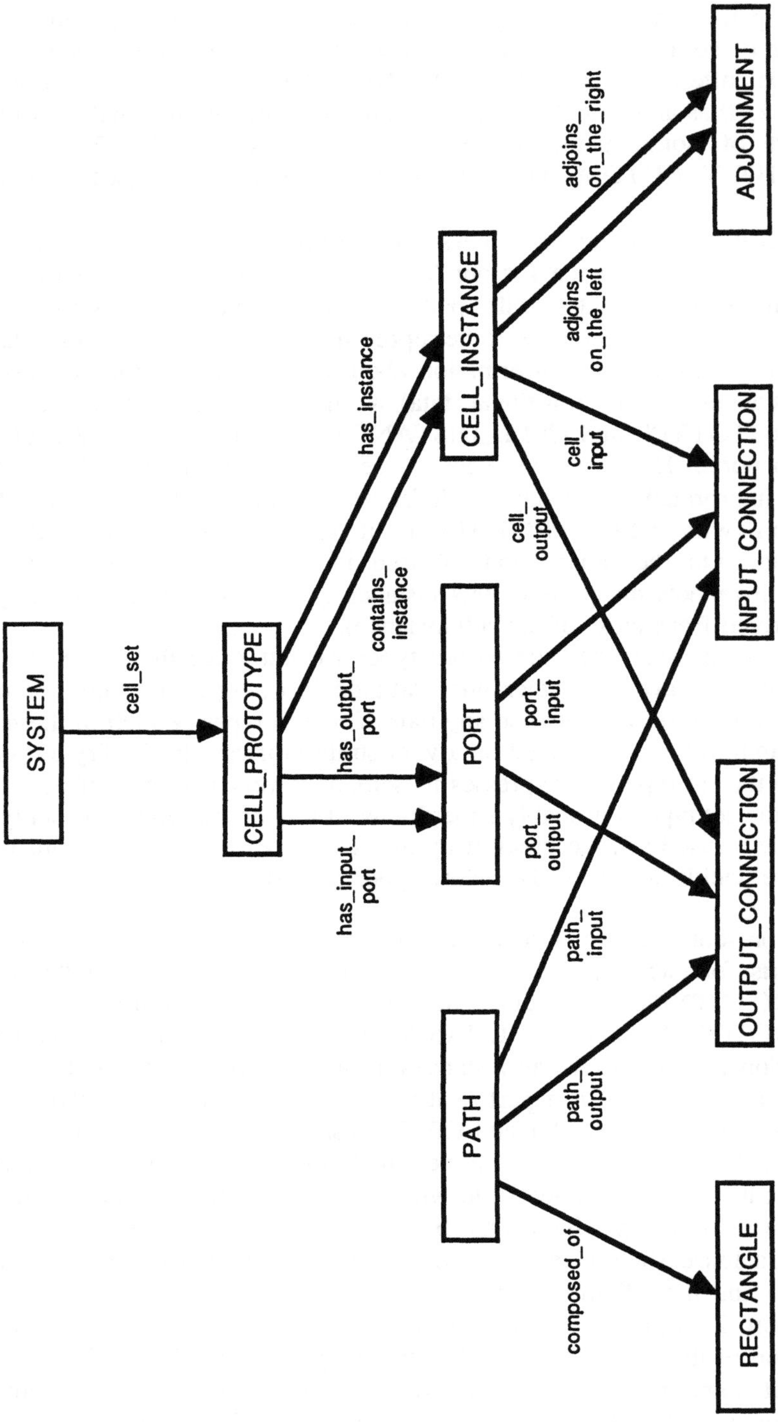

Fig. 2.16. Network schema for the design of Fig. 2.3

Which of the design rules from Sect. 2.1.4 are expressed in the schema of Fig. 2.16? The answer is not all that obvious and requires a careful analysis.

- Consider a cell instance. It may enter (via cell_output) into any number of output connections. In turn, each of these connections has only a single port (via port_output) and a single path (via path_output). This path may, of course, have other path_output members of type OUTPUT_CONNECTION, each with its own distinctive cell instance and port instance. Likewise, the path may be owner of any number of input connections. Consequently, no limitations have been imposed on paths; rule 1 and, by analogy, rules 2 and 3 remain unsatisfied.
- Rule 4 (except for the length restriction) is observed by the RECTANGLE record declaration.
- Rules 5, 6, and 7 cannot be enforced.
- Rule 8 is also not enforceable because we cannot express the fact that for a given CELL_PROTOTYPE the set of its has_instance members and the set of its contains_instance members must be disjoint.

The analysis is somewhat discouraging. Does it reflect an inherent weakness of the network model, or was it just clumsy design? There is a suspicion that it is the latter, since the first parts of rules 1 and 2 clearly establish a functional relationship from cell inputs to cell outputs. This relationship could easily be expressed by introducing port instances (that is, doing away with the original three-place relationship) and by distinguishing between input ports and output ports. Figure 2.18 shows the result (only the relevant portion of Fig. 2.16 has been redrawn). But unfortunately, now there is no

CELL_PROTOTYPE	(identifier: string; completion: date, released: boolean; technology: (CMOS, nMOS, ECL); function: string; size: integer).
PORT	(identifier: string; rel_x: integer; rel_y: integer).
CELL_INSTANCE	(identifier: string; x_position: integer; y_position: integer).
ADJOINMENT	( ).
PATH	(identifier: string; resistance: real; capacitance: real).
RECTANGLE	(identifier: string; length: integer; width: (wmin..wmax); x_origin: integer; y_origin: integer; orientation: integer).
INPUT_CONNECTION	( ).
OUTPUT_CONNECTION	( ).

Fig. 2.17. Details of the record types of Fig. 2.16

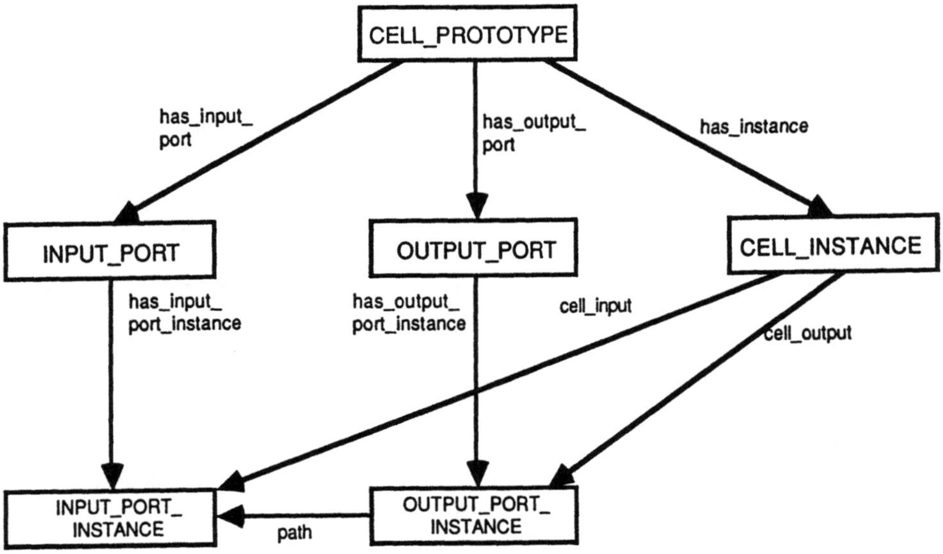

Fig. 2.18. An attempt to redesign the schema of Fig. 2.16

good way to maintain the path attributes. To do so would require replacement of the set type path by a link record, with a concomitant loss of the functional property.

We conclude: The network model is much better at reflecting the semantics of a miniworld, and it avoids redundancy. Still, it does not deal satisfactorily with what one might consider fairly simple constraints.

Since the network model is record-oriented, data manipulation is again by navigation. As a structurally much richer model it provides a considerably larger (and, one may add, more error-prone) variety of data manipulation options than the hierarchical model. We shall restrict ourselves to a brief summary of the more important features.

- Locating a record (*find*) is strictly separated from visibly transporting it into main memory. Hence, one may, e.g., delete a record without first moving it to main memory.
- Any record in the database may serve as an entrance record. It is advisable, however, to inform the DBMS, via the schema, of those records most frequently used, and the way they are identified to it (either by key or sequentially by system set are the two efficient options).
- Given a record, a number of options are available to navigate from there: If it is an owner, to the first member or one selected by key in a corresponding set; if it is a member, to the – in a predefined order – next member within the same or some other set, or to its owner. Also, it is possible to return to an unrelated record that was traversed before.
- All update operations, with the exception of storing a new record into the database, concern the most recent record reached by navigation. These

update operations are modifying or deleting a record, or inserting it into or removing it from a set.
- A new record is simply stored in the database. To implant it into a system set or some other set a separate operation is needed. However, if the AUTOMATIC option has been declared in the schema for the type of the new record vis-a-vis a certain set type, then storing the record will automatically add it as a member to a prespecified instance of that set type. If in addition the FIXED option has been used, then the only way to remove the record from that set is by deleting the record (for more details, see Sect. 2.2.2.3). Usually the two options go together, and are always specified for link record types. In the database of Fig. 2.16 the options also make sense for all cell instances in has_instance because instances must always have a prototype, but not for the cell instances in contains_instance because we may always encounter instances that are not part of other cells.

```
"Count the number of input ports for CELL 3"

count: = 0;
find    any CELL_PROTOTYPE using identifier = 'cell 3';
while  set_not_exhausted do
    find  next member PORT in has_input_port;
    count: = count + 1
endwhile;


"List all paths within CELL 3"

find    any CELL_PROTOTYPE using idenfier = 'cell 3';
while  set_not_exhausted do
    find  next member CELL_INSTANCE in contains_instance;
    while set_not-exhausted  do
          find next member INPUT_CONNECTION in cell_input;
          find owner PATH in path_input;
          get PATH;
          If this record has not yet been printed then print it
    endwhile;
    while set_not_exhausted do
          find next member OUTPUT_CONNECTION  in cell_output
          find owner PATH in path_output;
          get PATH;
          If this record has not yet been printed then print it
    endwhile
endwhile;


"Delete all input ports of CELL 3".

find any CELL_PROTOTYPE using identifier = 'cell 3';
while set_not_exhausted do
    find next member PORT in has_input_port;
    delete PORT
endwhile;
```

Fig. 2.19. Operations in a database satisfying the schema of Fig. 2.16

Figure 2.19 gives a few simple examples. As an exercise: What happens if one tries to delete the CELL3 prototype? To give a hint: Remember that cell instances make no sense in the database unless they have a corresponding prototype, and that instances may be part of other cells.

What we have given above is an outline of the "purist" network model which is being followed by many of the commercial DBMS products. Some of its rules have been relaxed over time. Already in its early form, the model permitted a less restrictive rule 4: Although the relationship still remains binary, more than one member record type may be declared for a given set type so that in essence a relationship is established between the owner record type and a union of record types. Rule 1 was changed by standards committee to read as follows: Owner and member record types may also be identical within a set type, although as a member the record must not be AUTOMATIC. In this case the same record can participate as a member in one set of the type and as an owner in another, or even as both owner and member within the same set. Consequently, hierarchies of records of a single type may now be constructed, e.g., Fig. 2.13a would be permissible. Finally, a few commercial products have also relaxed rule 3 and accept general (binary but otherwise unconstrained) relationships in which an owner still determines a set but a record may bexconnected as a member to more than one owner within the same set type. Because this kind of relaxation is costly in terms of implementation, only products with engineering applications in mind seem to have done so.

2.2.1.3 The Relational Data Model

The data manipulations in Fig. 2.19 demonstrate rather vividly that record-oriented DBMS interfaces require a lot of tedious programming. Consequently, they are helpful at best for database access from programs, and they are even naturally geared to that purpose because our programming languages can only deal with individual records rather than sets of records. For engineering applications this is no undue restraint because engineers very often access the database only indirectly via programs such as design tools or graphics packages.

Nevertheless, there are always good reasons for an engineer from time to time to inspect a database directly, and to do so interactively. Imagine querying the database for all paths within CELL3, and being forced to do so as in Fig. 2.19. The vendor of such a DBMS will hardly gain a new customer in this person!

As an alternative, we need a data model that deals not with individual records but with its sets. Such a model is called **set-oriented**. The next data model to be discussed is of this kind.

We start again with the structure concepts. First of all, since the model is not record-oriented, data manipulation by navigation does not make sense anymore so that only a single concept for representing the structural information is needed. Consider the set of instances of some record type and write it down as a table, with each row corresponding to an instance and each

CELL_PROTOTYPE

identifier	completion	released	technology	function	size
cell 1	851117	T	ECL	gate	560
cell 2	860120	T	ECL	gate	560
cell 3	860214	F	ECL	adder	12628

Fig. 2.20. A simple relation

column to a record attribute. Figure 2.20 demonstrates the idea for the cell prototypes. Any such table is called a **relation**, a column heading is a relation **attribute**, and a row a relation **tuple**. A relation is a set in the mathematical sense so that tuples remain unordered within a relation, and no duplicate tuples exist in a relation. Because of the closeness to mathematical relations, an extensive mathematical theory has been developed over the years for the relational model that has greatly contributed to the foundations of database technology in general. While we avoid a discussion of the theory in this book we shall refer to some of its results from time to time.

A relational schema consists of a list of relations, each represented by a name and a list of attributes together with their associated value sets (**domains**) much in the way shown in Figs. 2.5 and 2.17. In addition, if one attribute or a sequence of attributes are specified as **key** attributes of the relation, the values of a tuple under these attributes are supposed to identify the tuple uniquely within the relation. All relations are assumed to be in **first normal form (1Nf)**, i.e., all attribute values are atomic from the standpoint of the DBMS. For example, a completion date in CELL_PROTOTYPE (Fig. 2.17) can only be separated into year, month, and day outside the DBMS; the DBMS considers the date as a single unit and cannot perform, e.g., searches on the basis of year alone.

Although the concepts are simple, designing a "good" relational schema, i.e., one without redundancies, is by no means trivial. It is here that the relational theory comes in most handy, supporting a design process called the normalization of relations. Figure 2.21 gives one possible relational schema for the VLSI database. For each relation type defined in the schema, the database holds exactly one relation. This is illustrated in Fig. 2.22 with parts of a database corresponding to the schema Fig. 2.21.

The semantics of the schema become clearer if we compare it to the network schema Fig. 2.16. One simple rule is that any functional relationship (i.e., the inverse of a set) can be dealt with by adding an attribute to what originally was a member record type. For example, a cell instance has a unique type and a unique cell of which it is a part, hence we add attributes *prototype* and *envelope*. The corresponding CELL_PROTOTYPE tuples are identified within CELL_INSTANCE tuples by their CELL_PROTOTYPE keys; therefore, one refers to the attributes *prototype* and *envelope* as **foreign keys** in CELL_INSTANCE. Similar arguments hold for PORT and its foreign key attribute *prototype*; notice that the distinction between input and output

CELL_PROTOTYPE
(type_name: string; completion: date; released: boolean;
technology: (CMOS, nMOS, ECL); function: string; size: integer);
key type_name.

PORT
(port_name: string; prototype: string; mode: (I, O);
rel_x: integer; rel_y: integer);
key port_name.

CELL-INSTANCE
(instance_name: string; prototype: string;
envelope: string; x_position: integer; y_position: integer);
key instance_name.

ADJOINMENT
(left_cell: string; right_cell: string);
key (left_cell, right_cell).

RECTANGLE
(identifier: string; path: string; length: integer; width: (wmin..wmax);
x_origin: integer; y_origin: integer; orientation: integer);
key identifier.

CONNECTION
(path_name: string; out_cell: string;
out_port: string; in_cell: string;
in_port: string; resistance: real;
capacitance: real);
key path_name.

Fig. 2.21. Relational schema for the design of Fig. 2.3

ports has been relegated to the attribute *mode* although it would have been perfectly legal to introduce two identically structured relations INPUT_PORT and OUTPUT_PORT. Finally, RECTANGLE has been treated in the same fashion.

The same argument holds for ADJOINMENT with its foreign key attributes *left_cell* and *right_cell.* One could also argue less formally that ADJOINMENT was an artificial creation to begin with, and since the requirement of functionality does not hold here, we could directly translate Fig. 2.13a into a relation. This very argument was carried over to Fig. 2.14a and the relation CONNECTION.

Relation CONNECTION merits a more detailed analysis. First of all we note that, in accordance with our discussion of Fig. 2.16, port instances are uniquely identified by the combination of out_port (foreign key referring to the prototypical port) and out_cell (foreign key referring to the cell instance), and likewise of in_port and in_cell. We also note that, as a consequence, we should make sure that for any pair of in_port and in_cell the in_port value refers to a PORT tuple whose prototype value is identical to the prototype value in the CELL_INSTANCE tuple identified by the in_cell value. This shows how easy it is to build redundancy into a structurally simple data

CELL_PROTOTYPE

type_name	completion	released	technology	function	size
cell 1	851117	T	ECL	gate	560
cell 2	860120	T	ECL	gate	560
cell 3	860214	F	ECL	adder	12628

CELL_INSTANCE

instance_name	prototype	envelope	x_position	y_position
I 5	cell 2	cell 3	45	47
I 2	cell 1	cell 3	45	7
I 4	cell 1	cell 3	102	7
I 6	cell 3	NULL	0	0
I 3	cell 1	cell 3	75	7
I 1	cell 1	cell 3	18	7

ADJOINMENT

left_cell	right_cell
I 2	I 3
I 1	I 2
I 3	I 4

RECTANGLE

identifier	path	length	width	x_origin	y_origin	orientation
R 2	P 3	10	5	41	20	- 90
R 3	P 3	15	5	51	42	0
R 1	P 3	15	5	36	37	0

CONNECTION

path_name	out_cell	out_port	in_cell	in-port	resistance	capacitance
P 1	NULL	NULL	I 5	---	---	---
P 6	I 4	---	NULL	NULL	---	---
P 2	I 5	---	I 4	---	---	---
P 4	I 5	---	I 2	---	---	---
P 3	I 5	---	I 1	---	---	---
P 5	NULL	NULL	I 1	---	---	---

Fig. 2.22. Database for the example of Fig. 2.3 satisfying the schema of Fig. 2.21

a) **CELL_INSTANCE** (instance_name: string; prototype: string;
x_position: integer; y_position: integer);
key instance_name.

ENVELOPE (instance_name: string; envelope: string);
key instance_name.

b)

CELL_INSTANCE

instance_name	prototype	x_position	y_position
I5	cell 2	45	47
I2	cell 1	45	7
I4	cell 1	102	7
I6	cell 3	0	0
I3	cell 1	75	7
I1	cell 1	18	7

ENVELOPE

instance_name	envelope
I5	cell2
I1	cell1
I3	cell1
I4	cell1
I2	cell1

Fig. 2.23a,b. Redesign of relation CELL_INSTANCE from Fig. 2.21 to avoid NULL-values. **(a)** Schema. **(b)** Database

model. Finally, by choosing path_name as a key we fixed the interpretation of CONNECTION to one where a tuple identifies the path between a pair of ports (as opposed to Fig. 2.18 where a set represents the fork originating from an output port). These remarks should suffice to emphasize again the difficulties inherent in designing and interpreting a schema.

Relation CELL_INSTANCE raises another problem. If an enveloping cell does not exist, an undefined **(NULL)** value must be placed in the corresponding tuple position. Null-values have some unpleasant properties and can easily be avoided in this case by breaking down CELL_INSTANCE into two relations (Fig. 2.23). Cell instances that are not part of other cell types will appear in CELL_INSTANCE but not in ENVELOPE. On the surface, relation CONNECTION has the same problem. However, here the NULL values are indicative of a deeper problem: no distinction has been made between paths between cells and those from cells to envelope ports. We leave it to the reader as an exercise to split CONNECTION and to augment the resulting relations.

How is the topology of our example reflected in the database? Contrary to the first two models we have no concept that may be visualized graphically in the form of edges. A closer inspection of Fig. 2.22 gives the answer: Either it is directly embedded in a relation, or it is the foreign keys that provide the links between the tuples of different relations. The foreign keys could be considered as something akin to pointers – pointers that are not artificially created by the system but naturally by the user himself.

To summarize: The topology is not specified as such but is a matter of interpretation of the contents of the relation. This observation provides us with a clue as regards data manipulation. The database does not impose a certain topological view on the user; instead, he may freely choose any interpretation he wishes and at any time, as long as it can be based on the values within the relations. In this way, the user may even explore interdependencies that were not yet recognizable to him during problem analysis. Or to conclude from another perspective: Data manipulation in the relational model is not by structural analysis (navigation) but by content analysis (**associative access**).

Before we treat data manipulation in more detail, let us first examine the consequences of what we just said for the enforcement of the design rules from Sect. 2.1.4.

Rules 1 and 2 cannot be enforced because the system exerts no control over tuple values other than making sure that they are within a given domain. No functionality can be imposed in this way, and a general domain such as be exercised if one added a mechanism called **referential integrity:** Define as the domain of a foreign key the set of current keys in the corresponding relation.

For example, the domain of *left_cell* in ADJOINMENT would have to be CELL_INSTANCE [instance_name]. Note that this is the definition of a dynamic domain, i.e., one that changes over time. Note also that the mechanism implies that two cells may be joined in ADJOINMENT only *after* tuples were entered for them into CELL_INSTANCE. (Exercise: What happens if we try to delete a joined cell from CELL_INSTANCE?)

CELL-INSTANCE	(instance–name: string; prototype: CELL_PROTOTYPE [type_name]; envelope: CELL_PROTOTYPE [type_name]; x_position: integer; y_position: integer); **key** instance_name.
ADJOINMENT	(left_cell: CELL_INSTANCE [instance_name]; right_cell: CELL_INSTANCE [instance_name]); **key** (left_cell, right_cell).
CONNECTION	(path_name: string; out_cell: CELL_INSTANCE [instance_name]; out_port: PORT [port_name]; in_cell: CELL-INSTANCE [instance–name]; in_port: PORT [port_name]; resistance: real; capacitance: real; **key** path_name; **alternate key** in_cell:

Fig. 2.24. Expressing referential integrity and alternate keys in part of the schema of Fig. 2.21

The mechanism of referential integrity together with the separation of relation CONNECTION into two relations INPUT_PORT and OUTPUT_-PORT (rather than distinguishing the ports by mode) would then suffice to enforce the first part of rule 1.

Rule 2 cannot be enforced as the schema stands. By providing, together with referential integrity, a mechanism called the **alternate key** concept the problem may be overcome. An alternate key is a relation attribute that one may use in place of the key to identify a tuple uniquely. If we apply the concept to relation CONNECTION, declaring attribute *in_cell* to be an alternate key, the system guarantees that there is only a single path per cell input. Rules 3, 4 (as far as length is concerned), and 5 through 8 are unenforceable. Figure 2.24 summarizes the results.

Data manipulation, as mentioned before, is by associative means, and is set-oriented. By the latter term we mean that operations are applied to complete relations, and have new (complete) relations as their result. Since such operations define an algebra in mathematics, data manipulation in the relational model can be described in the form of a **relational algebra**. Again, we restrict ourselves to a brief introduction of the simpler operations.

- **Projection:** This operation, usually written R[A1,...,An] where n > 0, removes from relation R all columns except those for attributes A1,..,An. If the resulting relation contains duplicate tuples these are removed except for one. Notice that the dynamic domains above were declared in terms of projection.
- **Selection:** By contrast, this operation, written as R[A1,..,An $ v1,...,vn] where $ is a comparison operator and the vi are values, selects rows from a relation, namely those for which all comparisons A1 $ v1, ... , An $ vn are true.
- **Join:** This operation, written R[A1,...,An $ B1,...,Bn]S, is the one that allows following associative links between relations. It functions as follows. All those tuples of relations R and S are combined for which the $-comparisons hold for the values under A1 and B1, A2 and B2, and so on. If $ is the equality operator the columns under B1,...,Bn are omitted because they are redundant (**natural join**).

For an illustration of these operations see Fig. 2.25. Figure 2.26 demonstrates the application of an operation sequence. The reader is encouraged to construct the intermediate results after each step in order to convince himself that the result is correct. Incidentally, this is not the only way to construct the result. Can you find other sequences?

In fact, the relational algebra has never found favour with interactive users. Modern query languages such as SQL offer the user a way to express the properties of the desired result, leaving it to the DBMS to decide how to go about constructing it. Languages of this kind are often referred to as **descriptive languages** (or more pretentiously as fifth-generation languages). Figure 2.27 gives SQL expressions for the queries of Figs. 2.25 and 2.26.

"List all cell prototypes by name and function."

CELL_PROTOTYPE [type_name, function]

result

type_name	function
cell1	gate
cell2	gate
cell3	adder

"List all instances of cell 1."

CELL_INSTANCE [prototype = 'cell1']

result

instance_name	prototype	envelope	x_position	y_position
I 2	cell 1	cell 3	45	7
I 4	cell 1	cell 3	102	7
I 3	cell 1	cell 3	75	7
I 1	cell 1	cell 3	18	7

"List all cell instances that have a right adjoining cell, together with that cell."

CELL__INSTANCE [instance_name = left_cell] ADJOINMENT

result

instance_name	prototype	envelope	x_position	y_position	right_cell
I 2	cell 1	cell 3	45	7	I 3
I 3	cell 1	cell 3	75	7	I 4
I 1	cell 1	cell 3	18	7	I 2

Fig. 2.25. Simple relational algebraic operations on the database of Fig. 2.21

So far we have only dealt with retrieval of data. To update the database, one may use in the relational algebra set union and set difference, or in SQL special operations that handle sets of tuples. For the practical world we recommend updating one tuple at a time. That updates are not without their problems is illustrated in Fig. 2.28. Suppose we had been so careless as to pack the information on instances and prototypes into a single relation CELL. The information in CELL obviously is highly redundant. On updates we observe then the following **update anomalies:**

- **Insertion anomaly:** No new cell prototype can be introduced into the database without also introducing at least one instance of it.

"List all cell instances that have a right adjoining cell, by name, prototype and function."

```
T := CELL_INSTANCE [instance_name = left_cell] ADJOINMENT;
T := T [instance_name, prototype];
T := T [prototype = type_name] CELL_PROTOTYPE;
T := T [instance_name, prototype, function];
```

T

instance_name	prototype	function
I 2	cell 1	gate
I 3	cell 1	gate
I 1	cell 1	gate

Fig. 2.26. A sequence of relational algebraic operations on the database of Fig. 2.21. Whereas such a sequence is usually written as functional expression, we use here a programming language style, with T a relational variable

"List all cell prototypes by name and function."

```
select   type_name, function
from     CELL-PROTOTYPE
```

"List all instances of cell1."

```
select   *
from     CELL_INSTANCE
where    prototype = 'cell 1'
```

"List all cell instances that have a right adjoining cell together with that cell."

```
select   *
from     CELL-INSTANCE, ADJOINMENT
where    instance_name = left_cell
```

"List all cell instances that have a right adjoining cell, by name, prototype and function."

```
select   instance_name, prototype, function
from     CELL-INSTANCE, CELL-PROTOTYPE
where    instance_name in
         select   left_cell
         from     ADJOINMENT
```

Fig. 2.27. SQL expressions for the queries of Figs. 2.25 and 2.26

CELL

instance_name	type_name	technology	function	x_position	y_position
I5	cell 2	ECL	gate	45	47
I2	cell 1	ECL	gate	45	7
I4	cell 1	ECL	gate	102	7
I6	cell 3	ECL	adder	0	0
I3	cell 1	ECL	gate	75	7
I1	cell 1	ECL	gate	18	7

Fig. 2.28. A relation CELL constructed by joining CELL_PROTOTYPE and CELL_INSTANCE (only some of the attributes are shown)

- **Deletion anomaly:** If instance I6 is deleted from the database all information on prototype cell3 is also lost.
- **Replacement anomaly:** If the technology of cell1 is changed to CMOS, no fewer than four tuples in CELL must be modified.

The process of normalization mentioned earlier for designing relational schema provides the means for detecting such anomalies and splitting relations in an appropriate fashion.

2.2.2 Database System Interfaces

2.2.2.1 Database Languages and Standards

Data models are a set of concepts for structuring and manipulating databases. The engineer who is faced with a particular DBMS product, either while surveying the market, or because that is what the engineer's electronic data processing department is offering must determine:

- which data model the DBMS realizes;
- to what extent it realizes it; and
- which languages the DBMS offers for building and manipulating databases.

In this chapter we assume that a DBMS completely realizes a data model. We thus examine what typical languages look like for each of the three classical data models. Fortunately, quasi-standards seem to have been established in all three cases so that only older systems and some that do not quite fit any of the three models are not covered by the following discussion. The quasi-standards are: for the hierarchical model, the interface of IBM's system IMS (the only DBMS that realizes the hierarchical model); for the network model, the CODASYL Data Base Task Group reports of 1973 and 1978, also introduced at one time as a proposal for COBOL standardization; and for the relational model, the language SQL as one that is currently the subject of standardization efforts.

From a functional point of view we distinguish the following languages:

- **Data definition languages (DDL)** are for defining a database schema.
- **Data manipulation languages (DML)** are for database update and retrieval.
- **Query languages** are primarily intended for interactive access to a database. Often they are purely DML; however, sometimes they also contain elements of a DDL. Occasionally, query languages may also be used procedurally from application programs.
- **Procedural languages** are intended for procedural calls from programs. These languages are either wedded to particular programming languages (well-known examples are a network language for COBOL or a relational language for Pascal), or specified on a low (assembly language) level for wider applicability; in the latter case they are referred to as **CALL languages.** Like query languages, procedural languages are predominantly DML, with DDL elements being the exception.

2.2.2.2 The Hierarchical Data Model

The hierarchical model is probably the most widely used data model and has its strengths in business applications dealing with huge databases. Engineering applications have different requirements particularly as far as database sizes, work decentralization and data modeling capabilities go. The limitations of the latter were clearly demonstrated in Sect. 2.2.1.1. Because the hierarchical model is thus of little interest to the engineering user other than under exceptional circumstances, we only give the flavor of the system interface.

The schema of a hierarchical database consists of a number of database descriptions (DBD), one for each hierarchy. In turn, a DBD consists of the following statements:

- a DBD statement specifying, among other things, the name of the database;
- a DATASET statement that determines the database realization in terms of file and device type;
- a SEGM statement for each segment type in the hierarchical schema, with type name and parent name; and
- following each SEGM statement, one or more FIELD statements, one per segment field. One of the fields may be distinguished as defining the sort order of the segments in its class (SEQ) and is assumed to do so uniquely unless modified by 'M' ("multiple"). Further, physical information as to the relative byte position (START) and length of the field is given. Information on field domains is omitted because IMS does not check input values for validity.

Figure 2.29 illustrates this for the hierarchical schema of Fig. 2.7.

Strictly speaking, Figs. 2.7 and 2.29 represent a physical database. Suppose a schema is given and a database has been constructed according to the rules of Sect. 2.2.1.1. A database is called a **physical database** if all its segments are physically present on the storage media with all their fields. By contrast, a

```
DBD         NAME = VLSI_DESIGN
DATASET     DD = VLSIDD, DEVICE = 3370

SEGM        NAME = CELL_PROTOTYPE, BYTES = . . . . .
FIELD       NAME = (IDENTIFIER, SEQ), BYTES = 20, START = 1
FIELD       NAME = COMPLETION, BYTES = 6, START = 21
FIELD       NAME = RELEASED, BYTES = 1, START = 27
            o
            o
            o
            o
SEGM        NAME = PORT, PARENT = CELL_PROTOTYPE, BYTES . . . . .
FIELD       NAME = (IDENTIFIER, SEQ), BYTES = 20, START = 1
FIELD       NAME = INOUT, BYTES = 1, START = 21
            o
            o
            o
            o
SEGM        NAME = CELL_INSTANCE, PARENT = CELL_PROTOTYPE, . . . . .
FIELD       NAME = (IDENTIFIER, SEQ), . . . . .
            o
            o
            o
            o
SEGM        NAME = PATH, PARENT = CELL_PROTOTYPE, . . . . .
FIELD       NAME = (IDENTIFIER, SEQ), . . . . .
            o
            o
            o
            o
SEGM        NAME = RECTANGLE, PARENT = PATH, . . . . .
FIELD       NAME = (IDENTIFIER, SEQ),. . . . . .
FIELD       NAME = LENGTH, . . . . .
            o
            o
            o
            o
END         VLSI_DESIGN
```

Fig. 2.29. Database definition for the physical database of Fig. 2.7

database is called a **logical database** if, given a schema according to the rules of Sect. 2.2.1.1, the segments and fields associated with it either form a subset of some physical database, or are not all physically present, with those not physically present being constructed from physically present ones.

In terms of Sect. 2.1.2, the schema of a physical database corresponds to the conceptual schema, and the schema of a logical database to an external schema or subschema. Figure 2.30 shows a simple example of a subschema corresponding to a subset. The second situation is by far the more important one because it allows imposition of a network-like structure on the hierarchy and thus avoids redundancy. Since the concept is difficult to handle, inflexible, and incomplete, the engineer is well advised to turn to data models that are more natural to use.

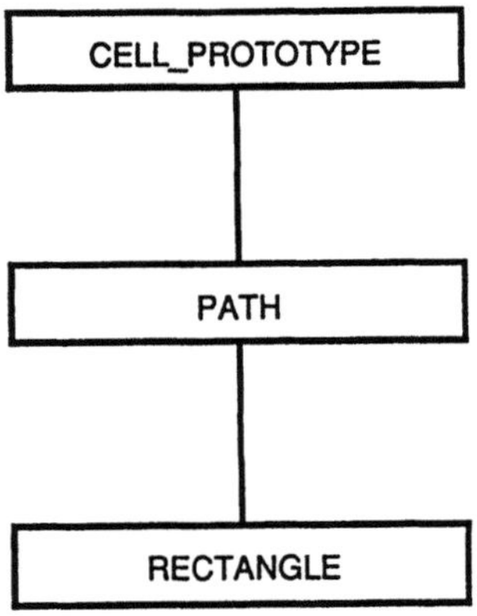

Fig. 2.30. A logical database derived from the physical database of Fig. 2.7

GU	get unique	Direct retrieval of a segment regardless of prior operations
GN	get next	Retrieval of the segment that follows the previously retrieved one in sequential order. Retrieval is restricted to a single class of some parent.
GNP	get next within parent	Retrieval of the segment that follows the previously retrieved one in sequential order. Retrieval reaches across all children of a given parent.
GHU GHN GHNP	get + hold	As GU, GN, GNP. To be used if a subsequent DLET or REPL is intended.
ISRT	insert	Add a new segment to a specified class.
DLET	delete	Delete the segment previously retrieved.
REPL	replace	Replace the segment previously retrieved.

Fig. 2.31. Summary of DL/1 operations

Data manipulation is strictly separate from data definition. The data manipulation language is procedural and thus a programmer's language. It was originally designed to be called from PL/1 programs, hence its name **DL/1**. Figure 2.31 summarizes the DL/1 operations. Since the communication between application programs and IMS is typical for all database systems, we give a brief outline for it. For reasons of process communication, IMS provides a single entry point (PLITDLI) rather than one for each DL/1 operation. Before calling IMS the application program must set up a work area into which or from which the segment of interest can be moved, and a communication area called the program communication area (PCB) where IMS leaves

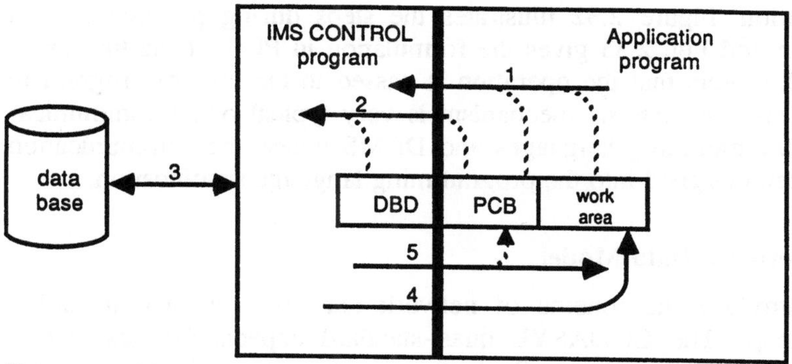

Steps:

1 Application program calls PLITDLI providing addresses of PCB and work area.
2 Control program interprets call on the basis of PCB and database definition DBD.
3 Control program searches and fetches segment from the database.
4 Control program transfers segment into work area.
5 Control program updates PCB and returns control to the application program.

Fig. 2.32. Steps during retrieval of a segment

```
DCL   1   VLSIPCB            BASED (VLSIPCB_ADDR),       /*PCB definition*/
          o
          o
          o
DCL   CELL2_WORK-AREA        CHAR (. .);                  /* input area */

DCL   1   CELL3SSA,                                 /* first-level expression */
      2   SEGNAME3           CHAR(8)    INITIAL ('CELL3'),
      2   LPAREN3            CHAR(1)    INITIAL ('('),
      2   FLDNAME3           CHAR(8)    INITIAL ('IDENTIFR'),
      2   COMPOP3            CHAR(1)    INITIAL ('='),
      2   FLDVAL3            CHAR(8)    INITIAL ('cell3'),
      2   RPAREN3            CHAR(1)    INITIAL(')');

DCL   1   CELL2SSA                                  /* second-level expression */
      2   SEGNAME2           CHAR(8)    INITIAL ('CELL2'),
      2   LPAREN2            CHAR(1)    INITIAL ('('),
      2   FLDNAME2           CHAR(8)    INITIAL ('IDENTIFR')
      2   COMPOP2            CHAR(1)    INITIAL ('='),
      2   FLDVAL2            CHAR(8)    INITIAL ('I5'),
      2   RPAREN2            CHAR(1)    INITIAL (')');

DCL   GU   CHAR(4)           INITIAL ('GU')           /* operation name */

DCL   FIVE  FIXED BIN (31)   INITIAL (5);             /* no of parameters */

CALL  PLITDLI  (FIVE, GU, VLSIPCB, CELL2_WORK_AREA,
                 CELL3SSA, CELL2SSA);                 /* call IMS */

IF   VLSIPCB.STATUS . . . . . . .. . . ;
```

Fig. 2.33. DL/1 program for the first example of Fig. 2.9

status information. Figure 2.32 illustrates the steps during processing of a DL/1 operation and Fig. 2.33 gives the formulation in PL/1 of the first operation in Fig. 2.9. Note that the operation is passed to the control program in string form. This cumbersome mechanism is very typical of all communication between programming languages and DBMS unless the communication has directly been designed into the programming language specification.

2.2.2.3 The Network Data Model

Before we introduce the schema of network databases we add a further structural concept. The CODASYL quasi-standard expects the designer to divide a database into one or more (logical) **areas**. In a sense, the area concept is a means for physical organization, suggesting which records are to be stored in physical proximity (e.g., on a given set of disks) so that they are accessible on-line all at once, or may simultaneously be moved from slower to faster peripheral storage or within a network. Areas are disjoint sets of records. The area concept is completely orthogonal to the record type and set concepts: An area may contain records of several types and several sets, but records of a given type or sets may also be spread across several areas.

A schema of a network database consists of the following sections.

- The first is an introductory SCHEMA statement which includes a schema name equal to the database name.
- An AREA section declares the areas into which the database is to be divided, by name and access characteristics.
- Next is a RECORD section in which all record types are defined. In turn, the definition of a record type consists of two subentries, one describing general characteristics such as type name, associated area(s), access authorization, identification within the area, and the other giving the record structure in terms of a hierarchical field arrangement.
- Finally, there is a SET section with a declaration for each set type. Again, such a declaration consists of two subentries. One outlines general properties such as type name, access authorization, owner type, and ordering criterion. The other subentry gives a detailed description of the properties of the member record with respect to a set instance of this type: member type, sort key, search keys, insertion and retention class, and rules for determining the set instance when a record of the member type is to be inserted into a set of this type.

The network DDL is highly complex, needing attention to many details. To avoid too much confusion we restrict ourselves in Fig. 2.34 to a simplified presentation of part of the database of Figs. 2.16 and 2.17. We briefly discuss some of its aspects.

Only a single area has been defined. The RECORD declarations limit themselves to type name, associated area (WITHIN; trivially the only one specified), identification within the area (KEY; if DUPLICATES NOT ALLOWED is specified the key is unique within the area), and record structure. Records of types with a KEY clause may serve as entrance points into the

```
SET NAME IS CELL_SET;
    OWNER IS SYSTEM;
    ORDER IS SORTED BY DEFINED KEYS
        DUPLICATES ARE NOT ALLOWED.
    MEMBER IS CELL_PROTOTYPE;
        INSERTION IS AUTOMATIC
        RETENTION IS MANDATORY;
        KEY IS ASCENDING IDENTIFIER IN CELL_PROTOTYPE;
        SET SELECTION IS BY SYSTEM.

SET NAME IS HAS_INPUT_PORT;
    OWNER IS CELL_PROTOTYPE;
    ORDER IS LAST;
    MEMBER IS PORT;
        INSERTION IS AUTOMATIC
        RETENTION IS FIXED;
        SET SELECTION IS BY VALUE OF IDENTIFIER IN CELL_PROTOTYPE.

SET NAME IS HAS_OUTPUT_PORT;
    o
    o
    o

SET NAME IS HAS_INSTANCE;
    OWNER IS CELL_PROTOTYPE;
    ORDER IS LAST;
    MEMBER IS CELL_INSTANCE;
        INSERTION IS AUTOMATIC
        RETENTION IS FIXED;
        SET SELECTION IS BY VALUE OF IDENTIFIER IN CELL_PROTOTYPE.

SET NAME IS CONTAINS_INSTANCE;
    OWNER IS CELL_PROTOTYPE;
    ORDER IS LAST;
    MEMBER IS CELL_INSTANCE;
        INSERTION IS MANUAL
        RETENTION IS FIXED;
        SET SELECTION IS BY CURRENT OF OWNER.
    o
    o
    o
```

Fig. 2.34. Part of a CODASYL schema for the database of Figs. 2.16 and 2.17

database. Note that link records such as ADJOINMENT do not have a key clause; they also have no fields.

The SET declarations (only very few are shown) contain type name, order criterion, member type, insertion and retention class, and set selection clause. Two order criteria are demonstrated, one that sorts by the key defined in the member subentry, and one that appends to the end of the set (LAST). The insertion class determines whether on storing a record of the member type into the database it should automatically be included in some instance of the set type (AUTOMATIC; the alternative is MANUAL). For example, ports have no existence of their own but depend on cell prototypes. Likewise, each cell instance is of some prototype, however, an instance is not necessarily included in some cell prototype. On the other hand, the retention class determines whether a record of the member type may be removed from an instance of the set type without deleting it at the same time from the database (OPTIONAL), or only by simultaneously deleting it (FIXED), or by simultaneously connecting it to some other instance of the same set type (MANDATORY). Note that in general AUTOMATIC and MANDATORY or FIXED go together. However, in case of CONTAINS_INSTANCE a cell instance does not necessarily go into a prototype; if it does, though, it has lost its independence. Finally, set selection determines how, in case of an AUTOMATIC insertion of a record of the member type, the instance of the set type is to be chosen. This is done by selecting the corresponding owner either by some identifying value (BY VALUE), or by its currency indicator (BY CURRENT, see below), or trivially in case of a system set (BY SYSTEM).

We note that a CODASYL schema does not limit itself to structural aspects but also includes descriptive elements for controlling the detailed behavior of the update procedures.

As a logical consequence of following the three-schema approach, application programs do not deal with the schema itself but with a subschema. It may differ from the schema in several respects.

- Record and set types may be missing.
- Record fields may be missing, their domains may be changed, and the fields may be renamed or arranged in different hierarchical order.
- The set selection clause may be overridden by a new one.
- The subschema DDL is not neutral but conforms to the syntax of some higher programming language (implying, incidentally, that it is integrated into the specification of that language, e.g., COBOL).

Figure 2.35 gives a simple example.

```
SCHEMA NAME IS VLSI_DESIGN.

AREA NAME IS VLSI_AREA.

RECORD NAME IS CELL_PROTOTYPE;
        WITHIN VLSI_AREA;
        KEY IS IDENTIFIER DUPLICATES ARE NOT ALLOWED.
                IDENTIFIER;   TYPE IS CHARACTER 20.
                COMPLETION;   TYPE IS CHARACTER  6.
                RELEASED;     TYPE IS CHARACTER  1.
                o
                o
                o
RECORD NAME IS PORT;
        WITHIN VLSI_AREA;
        KEY IS IDENTIFIER DUPLICATES ARE NOT ALLOWED.
                IDENTIFIER;   TYPE IS CHARACTER     20.
                RELX;         TYPE IS FIXED DECIMAL 3.
                RELY;         TYPE IS FIXED DECIMAL 3.
```

Fig. 2.35 (continued)

```
RECORD NAME IS CELL_INSTANCE;
       WITHIN VLSI_AREA;
       KEY IS IDENTIFIER DUPLICATES ARE NOT ALLOWED.
              IDENTIFIER;   TYPE IS CHARACTER 10.
              XPOS;         TYPE IS FIXED DECIMAL 3.
              YPOS;         TYPE IS FIXED DECIMAL 3.

RECORD NAME IS ADJOINMENT;
       WITHIN VLSI_AREA.

RECORD NAME IS INPUT_CONNECTION;
       WITHIN VLSI_AREA,
              o
              o
              o
RECORD NAME IS OUTPUT_CONNECTION;
              o
              o
              o
RECORD NAME IS IS PATH;
              o
              o
              o
RECORD NAME IS RECTANGLE;
              o
              o

TITLE DIVISION.

SS   CELL_PORTS WITHIN VLSI_DESIGN.

STRUCTURE DIVISION.

REALM SECTION.
RD   CELL_REALM CONTAINS CELL_PROTOTYPE, PORT RECORDS.

RECORD SECTION.
01   CELL_PROTOTYPE.
     02 PROTONAME;        PICTURE IS X(20).
     02 COMPLETION.
        03 DATE;          PICTURE IS X(6).
        03 RELEASED;      PICTURE IS X(1).
     o
     o
     o
01   PORT.
     02 PORTNAME;         PICTURE IS X(20).
     02 RELPOS.
        03 X;             PICTURE IS S9(3).
        03 Y;             PICTURE IS S9(3).

SET SECTION.
SD   HAS_INPUT_PORT.
SD   HAS_OUTPUT_PORT.
```

Fig. 2.35. COBOL subschema derived from the schema of Fig. 2.34

As mentioned earlier, data manipulation is organized much in the way of Fig. 2.32. The data manipulation language is considerably more complex than that for the hierarchical model due to the structural richness and the navigational variety in the network model. As pointed out in Sect. 2.2.1.2, all operations apply to the record reached last during the navigation. It is referred to as the **current_of_run_unit**, and the DBMS maintains an internal pointer with the same name. In order to return to records reached earlier during navigation (e.g., in nested loops such as in the second example of Fig. 2.19),

FIND ANY record_name1 USING field_name IN record_name2
Entering the database via some record of type record_name1 identified by the value of field field_name in the work area for record_name2. Access is most efficient if record_name2 equals record_name1 and field_name is the area key.

FIND DUPLICATE record_name1 USING field_name IN record_name1
Navigating to the next record in the area with the same value in field_name.

FIND FIRST record_name WITHIN area_name
FIND NEXT record_name WITHIN area_name
Step-by-step sequential access to all records of type record_name in area area_name.

FIND FIRST record_name WITHIN set_name
FIND NEXT record_name WITHIN set_name
Step-by-step sequential access to all members of the set of type set_name identified by the current_of_run_unit.

FIND record_name1 WITHIN set_name CURRENT
USING field_name IN record_name2
Direct access to the member of type record_name1 identified by the contents of field_name in work area record_name2. The set instance concerned is the one given by the contents of current_of_set_name. Access is most efficient if record_name2 equals record_name1 and field_name is search key. CURRENT may be omitted; in this case the set instance is determined by interpretation of the set selection clause of record_name1 in set_name.

FIND DUPLICATE WITHIN set_name USING field_name IN record_name
Continuation from the record located in the previously discussed operation, where the values for field_name are the same.

FIND OWNER WITHIN set_name
Locates the set_name_owner to the current_of_run_unit.

FIND CURRENT record_name WITHIN area_name
FIND CURRENT record_name WITHIN set_name
Makes the current_of_area_name, or the current_of_set_name, or (if WITHIN has been omitted) the current-of_record_name the current_of_run_unit.

Fig. 2.36. Navigation operators in the network model

additional pointer variables must be retained. A certain set of these (**currency indicators**) is maintained by the DBMS. Besides the current_of_run_unit these are: a pointer to the last record of each record type reached during navigation (**current_of_record_type**), a pointer to the last record reached in each area (**current_of_area**), and a pointer to the last record reached in some instance of each set type (**current_of_set_type**; note that this also identifies a set instance). These pointers are automatically updated with each operation. For example, if we navigate from a cell prototype to a cell instance, the following currency indicators are modified to the CELL_INSTANCE record reached:

CURRENT_OF_RUN_UNIT,
CURRENT_OF_VLSI_AREA,
CURRENT_OF_CELL_INSTANCE,
CURRENT_OF_HAS_INSTANCE,
CURRENT_OF_CONTAINS_INSTANCE,
CURRENT_OF_ADJOINS_ON_THE_RIGHT,
CURRENT_OF_ADJOINS_THE_LEFT,
CURRENT_OF_CELL_INPUT, and
CURRENT_OF_CELL_OUTPUT.

GET record_name
Retrieves the current_of_run_unit into the work area.

ERASE record_name
In its simplest form, deletes the current_of_run_unit from the database, and hence from all sets of which it currently is a member.

STORE record_name
Inserts the record in work area record_name into the database.
For all set types in which record_name is an AUTOMATIC member, set instances are determined according to the respective set selection clauses, and the new record is inserted into them according to the order criterion.

MODIFY record_name
Updates the current_of_run_unit.

CONNECT record_name TO set_name
Inserts the current_of_run_unit into the instance of set type set_name determined by the set selection clause for record_name. Insertion class must be MANUAL.

DISCONNNECT record_name FROM set_name
Removes the current_of_run_unit from its set instance of set type set_name. Retention class must be OPTIONAL.

RECONNECT record_name WITHIN set_name
Switches the current_of_run_unit from its present set instance of set type set_name to a new instance determined by the corresponding set selection clause.

Fig. 2.37. Retrieval and update operators in the network model

There are two ways to circumvent the problems this may cause when the pointer to a record one would like to return to gets wiped out: RETAIN the values of specified indicators, or KEEP these indicators' values in a list controlled by the application program (the DML includes the appropriate language elements). The novice should be warned that the concept of the currency indicators is a frequent source of program errors.

Like the subschema DDL, the DML is defined to be part of the specification of some programming language. The ANSI COBOL standard of 1976 includes a proposal for access to network databases. Figures 2.36 and 2.37 summarize the DML operators in the syntactic form in which they may be used in COBOL programs.

2.2.2.4 The Relational Data Model

At least in its interactive version with SQL as the quasi-standard, the relational model does not strictly separate data definition from data manipulation. For the declaration of relations (more precisely: of those relations that are to be

```
CREATE  TABLE CELL_PROTOTYPE
        (type_name          (CHAR (20), NONULL),
         completion         (CHAR ( 6)),
         released           (CHAR ( 1), NONULL),
         technology         (CHAR ( 4), NONULL),
         function           (CHAR (10), NONULL),
         size               (INTEGER)  )

CREATE  UNIQUE INDEX TNAMEX  ON CELL_PROTOTYPE (type_name)

CREATE  TABLE PORT
        (port_name          (CHAR (20), NONULL),
         prototype          (CHAR (20), NONULL),
         mode               (CHAR ( 1), NONULL),
         rel_x              (INTEGER),
         rel_y              (INTEGER)  )

CREATE  UNIQUE INDEX CNAMEX ON PORT (port_name)

CREATE  TABLE CELL_INSTANCE
        (instance_name      (CHAR (20), NONULL),
         prototype          (CHAR (20), NONULL),
         envelope           (CHAR (20)),
         x_pos              (INTEGER),
         y_pos              (INTEGER)  )

CREATE  UNIQUE INDEX INAMEX ON CELL_INSTANCE (instance_name)

CREATE  TABLE ADJOINMENT
        (left_cell          (CHAR (20), NONULL),
         right_cell         (CHAR (20), NONULL))

CREATE  UNIQUE INDEX ADJX ON ADJOINMENT (left_cell, right_cell)
```

Fig. 2.38. Part of an SQL dialogue for declaring the schema of Fig. 2.21

physically represented in the database, the **base relations**) a CREATE TABLE operation exists. (In fact, one may easily expand a table by further attributes later on using an EXPAND TABLE operation.) Keys cannot be specified directly. Instead, one may associate an index with any attribute for faster access, and mark it as unique, by a CREATE INDEX operation. Figure 2.38 shows what a dialog may look like when one declares the schema of Fig. 2.21. Relations including all indexes associated with them may be dropped by DROP TABLE. Note that by marking an attribute with the NONULL option the DBMS is advised not to accept tuples with an undefined value for this attribute.

The relational model offers a more powerful approach to logical databases than either of the other two classical models. Basically, any logical relation may be formed that can be expressed as the result of a SELECT operation. Such a relation is called a **derived relation** or a **view**. Views may be dynamically created (DEFINE VIEW) and removed (DROP VIEW). Views are not physically kept in the database but are constructed from the base relations according to their SELECT expressions whenever they are accessed, hence they reflect the latest changes to the base relations. Views may be subjected to retrieval operations (SELECT) like any base relation. Modifications to views, however, often cannot be translated into corresponding changes to base relations and, consequently, should in general be avoided. A view may be transformed into a permanent relation by the KEEP TABLE operation. Figure 2.39 illustrates the view definition for the example of Fig. 2.28.

The operations for data manipulation are summarized in Fig. 2.40. Examples for retrieval have already been treated in Fig. 2.27, although much more complex queries are possible. The update operations are also set-oriented, i.e., sets of tuples may be affected by a single update operation, with

```
DEFINE VIEW CELL AS

     SELECT  instance_name, type_name, technology, function,
             x_position, y_position
     FROM    CELL-PROTOTYPE, CELL_INSTANCE
     WHERE   type_name = prototype
```

Fig. 2.39. Definition of the view given in Fig. 2.28

SELECT	retrieve elements from the database and assemble them into a relation
UPDATE	modify the values in a relation
INSERT	insert new tuples into a relation
DELETE	delete tuples from a relation

Fig. 2.40. SQL data manipulation operators

"Move all cell1 instances in cell3 to the left by 5"

```
UPDATE      CELL_INSTANCE
SET         x-position = x_position - 5
WHERE       prototype = 'cell1'
AND         envelope = 'cell3'
```

"Remove path P3 from the database"

```
DELETE      RECTANGLE
WHERE       path = 'P3'

DELETE      CONNECTION
WHERE       path_name = 'P3'
```

Fig. 2.41. Examples of multiple-tuple updates in the database of Figs. 2.21 and 2.22

```
$DCL        SHIFT   FIXED BIN (31);
 DCL        ALPHA ........;
 DCL        BETA  ........;
  o
  o
  o
 IF ALPHA > BETA
$UPDATE     CELL_INSTANCE
  SET       x_position = x_position - $SHIFT
  WHERE     prototype = 'cell1'
  AND       envelope = 'cell3';
  o
  o
  o
```

Fig. 2.42. Example of an SQL statement embedded in PL/1. Note that PL/1 variables are referenced within an SQL expression by a $-prefix, and their declaration is also prefixed by $

the single-tuple update the special case. Figure 2.41 gives two examples. (Note that in the second example the first DELETE is a single-tuple operation, the second is multiple-tuple. Why?)

Of particular interest to application programs is whether they may also use SQL to access a relational database. Indeed, the approach taken is exactly the same as discussed in Sect. 2.2.2.2: Provide an entry point into the DBMS, and call the associated procedure with the SQL expression passed as a string parameter. An even more comfortable solution avoids the translation of the expression into string parameters by the programmer and leaves it to a precompiler, permitting the use of the SQL syntax in the middle of the programs. SQL expressions must just be marked as such so that they can be recognized by the precompiler. The greatest advantage of this approach is that one may use program variables within the SQL expressions. Figure 2.42 gives a simple example for PL/1.

```
$DCL     TN     CHAR (20);
$DCL     FU     CHAR (10);

$LET CU BE
  SELECT   type_name, function
  INTO     $TN, $FU
  FROM     CELL_PROTOTYPE;

$OPEN CU;
  DO WHILE more_to_come;
     $FETCH CU
    o
    o
    o
```

Fig. 2.43. Use of the cursor concept

There remains one mismatch, however: SQL is set-oriented, programming languages are record-oriented. The mismatch is resolved by not retrieving a set directly into the program but into something akin to a sequential file, called a **cursor**. In order to process the tuples after they have been retrieved the cursor must be opened, and tuples fetched sequentially, one at a time. Figure 2.43 demonstrates the use of a cursor for the query at the top of Fig. 2.27.

2.2.3 Database Consistency

2.2.3.1 Consistency Control

Generally speaking, a database is consistent if at any time it is a faithful representation of some predetermined part of the application world (the **miniworld**). Obviously, no DBMS has any chance of guaranteeing consistency in that general sense. To give an example, how should it recognize that a user marks a cell prototype as released even though it is still under work? Clearly then, whenever the DBMS is not a direct observer of the miniworld but works through the user as an interlocutor, only a weaker notion of consistency can apply: A database is **consistent** if at any time it satisfies an explicitly specified set of verifiable rules.

Consequently, consistency is not some absolute property of a database but has always to be seen in light of a given set of rules. The more specific and voluminous the set is, the closer consistency comes to the ideal stated in the beginning.

The rules evolve from an analysis of the miniworld. A point in case are the design rules of Sect. 2.1.4. Of these, very few could be directly reflected in the schema using the basic concepts of the data model. Which ones were and which ones were not depended on the data model chosen. All those that could not be incorporated had, by necessity, to be ignored and, hence, were outside the set of rules that defined the data model inherent consistency of the database.

Consequently, any realistic goal of database consistency requires as part of the DBMS a mechanism for expressing rules that cannot be taken care of by the chosen model, and for controlling the execution of a database operation such that it leaves no trace in the database if at least one of the rules is being contradicted as the result of the operation (**consistency control**). For example, if one added a new path to a cell design, the mechanism would inspect the result of the update operations involved to ascertain that rule 6 is being met, i.e., that the new path does not intersect an existing one. If it would, the update is rejected. Rules that are expressed to the consistency control and are enforced by it are called **consistency constraints.**

As a consequence of what was said before, there is nothing absolute in the need for and scope of the consistency constraints; these depend on the power of the data model. For example, if we use a relational DDL that includes constructs for referential integrity and alternate keys, then the data model and, hence, the relational operators take care of parts of rules 1 and 2. If, on the other hand, they are not included, we require the consistency control mechanism to which we express our needs in the form of additional consistency constraints. Figure 2.44 depicts the ensuing architecure.

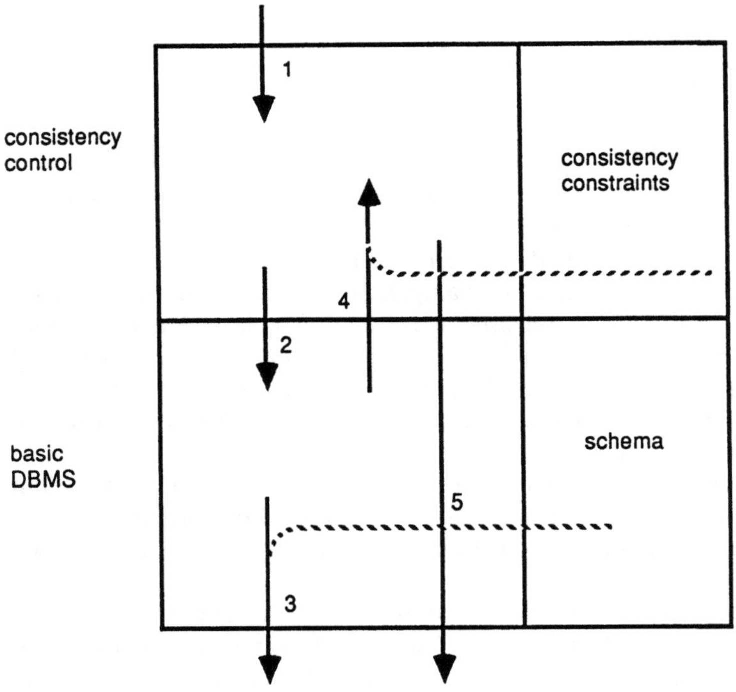

Fig. 2.44. Logical architecture of a DBMS with consistency control and data and control flow for database updates. 1: update request to DBMS; 2: unchanged update request to basic DBMS; 3: processing of update and temporary storage of result; 4: consistency check of result; 5: if result is consistent, permanent storage of result

```
a) begin transaction
       find any CELL_PROTOTYPE using identifier = 'cell3';
       modify CELL_PROTOTYPE field technology to 'CMOS';
       while set_not_exhausted do
              find next CELL_INSTANCE within contains_instance;
              find owner within has_instance retaining contains_instance;
              modify CELL_PROTOTYPE field technology to 'CMOS'
       endwhile
   end transaction;
```

```
b) update   CELL_PROTOTYPE
   set      technology = 'CMOS'
   where    type_name = 'cell3'
   or       type_name in
             select   prototype
             from     CELL_INSTANCE
             where    envelope = 'cell3'
```

Fig. 2.45a,b. Update query for "Change the technology of all cells of the CELL3 design to CMOS". **(a)** Network query based on the schema of Fig. 2.16. The DML has been somewhat simplified compared to Figs. 2.36 and 2.37. Note that in the form given the transaction may perform some unnecessary updates. **(b)** Relational query based on the schema of Fig. 2.21

Given a set of constraints and a certain database state, it is quite possible that a desired new database state cannot be reached in a single update step. Take design rule 6 as a constraint and suppose, e.g., that the cell technology in the CELL3 design is to be changed from ECL to CMOS. Clearly, in a network database the appropriate attribute would have to be changed in three records, necessitating a number of database accesses and update steps (Fig. 2.45). Only after the last update step has been completed is the database again consistent. On the other hand, a set-oriented update in the relational model performs the same transformation in a single step, without intermittent inconsistencies (Fig. 2.45).

For such a multiple-step transformation one would like to modify the consistency control such that it does not check the consistency after each step but postpones the check until after the last step has been completed. This is done by bracketing the sequence of steps by BEGIN TRANSACTION and END TRANSACTION (Fig. 2.45). The sequence itself is called a **database transaction.** A single update step outside these brackets could then be considered a "trivial" transaction.

A transaction is thus a "unit of consistency". This suggests a different point of view: By bracketing the sequence we notify the DBMS of the fact that the database is not to be considered consistent unless all steps have been completed, no matter whether intermediate results within the sequence meet the consistency constraints or not. In other words, only the sequence as a whole reflects an action in the miniworld, or, to put it in the negative: The constraints do not sufficiently reflect the laws of the miniworld. Figure 2.41 (second example) is such a case and must be formulated as in Fig. 2.46. (The

```
BEGIN TRANSACTION

      DELETE     RECTANGLE
      WHERE      path = 'P3'

      DELETE     CONNECTION
      WHERE      path_name = 'P3'

END TRANSACTION
```

Fig. 2.46. The example of Fig. 2.41 as a transaction

intermediate state is consistent no matter whether referential integrity is observed or not. Why?)

Consistency constraints can be classified according to several criteria.

1. **Constraint granularity.** A constraint establishes a relationship between one or more data objects. In the database of Fig. 2.21 the relationship may range from a single attribute ("the technology values can only be CMOS, ECL or nMOS"), to several attributes within a record ("prototype and envelope in CELL_INSTANCE cannot be the same", a rephrasing of "a cell cannot contain a cell of its own type"), to several records of the same type ("no cells may overlap"), and to several records of different types ("the in_port of a connection must have mode = I").
2. **Consistency delay.** Consistency checks may be delayed. If they are not, consistency is enforced after every update operation. The typical mechanism for delaying consistency checking is the transaction concept introduced above.
3. **Constraint type.** A consistency constraint may express static or dynamic properties. Static constraints are those that apply to a database state, and may be enforced by inspecting the database in a fixed state. Dynamic constraints describe legal transitions from one database state to another, and can only be verified by taking the operations into account. All constraints given so far are static. A dynamic constraint could be "a design in CMOS may only be changed to ECL".

2.2.3.2 Expressing Consistency Constraints

Consistency enforcement is a rather expensive affair. Consequently, few of today's systems offer more than rudimentary facilities for it. We briefly examine the constructs that have been proposed for the network and relational languages.

The network model contains two constructs that may be used for expressing consistency constraints. The first is a CHECK clause for undelayed static constraints that apply to single attributes, several attributes within a record, or records within a set. Figure 2.47 gives some examples. Note that the first of these is due to the fact that the network operations as such do not work with enumeration types as domains.

a) Part of the declaration of field TECHNOLOGY in CELL_PROTOTYPE:

```
CHECK IS VALUE 'CMOS', ' ECL', 'nMOS'
```

b) Part of the declaration of field WIDTH in RECTANGLE:

```
CHECK IS VALUE wmin THRU wmax
```

c) Part of the declaration of record RECTANGLE:

```
CHECK IS LENGTH > WIDTH
```

d) Part of the declaration of set HAS_INSTANCE

```
CHECK IS PROTOTYPE IN CELL_INSTANCE = IDENTIFIER IN
CELL_PROTOTYPE
```

Fig. 2.47a–d. Consistency constraints in the database of Figs. 2.16 and 2.17 using the CHECK clause. **(a)** Single attribute-based. **(b)** Single attribute-based. **(c)** Record-based. **(d)** Set-based; record CELL_INSTANCE is assumed to have been extended by a field PROTOTYPE

We note also that we still do not account for design rules 5 through 8 of Sect. 2.1.4. Apparently the CHECK mechanism is not powerful enough. With the ON clause the network language provides an escape by allowing the association of user-written test procedures with specific update operations. These procedures are automatically invoked at the end of the update. Suppose that the test for design rule 7 is encoded as a procedure CELL_OVERLAP_-TEST. Figure 2.48 illustrates how this procedure is connected to the record definition of CELL_INSTANCE.

```
ON STORE CALL CELL_OVERLAP_TEST
```

Fig. 2.48. Procedurally defined consistency constraint in thew declaration of record type CELL_INSTANCE that is checked after STOREing a cell instance

In its original design the relational language SQL included two consistency control facilities as well, assertions and triggers. An assertion describes a constraint that is to be checked and enforced at the end of one of a specified class of update operations, e.g., after each modification to one or more tuples of a particular relation. Assertions can be defined at all levels of granularity; and they may express static or dynamic constraints. Figure 2.49 shows several examples. Assertions also allow undelayed or delayed checks: If the update operations are part of a transaction the check is postponed to the end of the transaction, unless the assertion is marked IMMEDIATE to indicate enforcement without delay, or the associated constraint is dynamic.

Triggers are a generalization of the concept of assertion. A trigger causes a prespecified sequence of SQL statements to be executed whenever some

```
a)    ASSERT A1 ON CELL_PROTOTYPE:
              TECHNOLOGY IN ('CMOS', ' ECL', 'nMOS')

b)    ASSERT A2 ON INSERTION OF CELL_INSTANCE,
                     UPDATE OF CELL-INSTANCE:
              PROTOTYPE ≠ ENVELOPE

c)    ASSERT A3 ON INSERTION OF CELL_INSTANCE:
              (SELECT TECHNOLOGY FROM NEW CELL_INSTANCE)
              IS IN
              (SELECT TECHNOLOGY FROM OLD CELL_INSTANCE)

d)    ASSERT A4 ON UPDATE OF CELL_INSTANCE:
              NEW TECHNOLOGY = OLD TECHNOLOGY
```

Fig. 2.49a–d. Assertions for the database of Fig. 2.21. **(a)** Single attribute-based, static. **(b)** Multiple attribute-based, static; the assertion reflects design rule 8. **(c)** Relation-based, dynamic; the assertion enforces design rule 6. **(d)** Single attribute-based, dynamic; the assertion completes enforcement of design rule 6

```
DEFINE TRIGGER T1

      ON DELETE OF CONNECTION:
              (DELETE RECTANGLE
              WHERE RECTANGLE . path = CONNECTION . path_name)
```

Fig. 2.50. The example of Fig. 2.41 as a trigger

triggering event occurs. The triggering event may be retrieval, insertion, deletion or update of a particular relation or even view. The granularity, delay, and type properties are identical to those of assertions. Figure 2.50 illustrates the trigger concept for the Qarlier transaction example (Fig. 2.46). Note that this appears to be the most natural formulation because the user usually just wants to drop the connection, without having to worry about the various places where information on the connection is kept, and where referential integrity must be observed.

Like relations, views, and indexes, assertions and triggers are supposed to be dynamically created (ASSERT, DEFINE TRIGGER) and deleted (DROP, in this case they must have been named). In order to express enforcement delays, transactions must be defined; they are bracketed by BEGIN_TRANS and END_TRANS.

2.2.4 Concurrency

Databases are a technical means for integrating the data requirements of many persons in an enterprise, and an organizational means for coordinating their activities. Consequently, one will have to expect that a fair number of users will concurrently access and manipulate the database. Even worse, they may sometimes access the same data, and in this way interact involuntarily and unconsciously. The objective of **concurrency control** in a DBMS is to avoid such interaction while at the same time maintaining the maximum degree of parallelism between the user processes.

What kind of interactions may occur? To arrive at an answer let us study a few examples. Consider the relational database of Fig. 2.22 (schema in Fig. 2.21). Now suppose that four different queries have been submitted to the DBMS to be executed more or less simultaneously (Fig. 2.51; some of these queries could be formulated somewhat more intelligently, but for the sake of illustration we leave them as they are).

P1 : "Change the technology of all cells that have to do with CELL3 to CMOS."

```
(1)  UPDATE   CELL_PROTOTYPE
     SET      technology = 'CMOS'
     WHERE    type_name = 'cell1'

(2)  UPDATE   CELL_PROTOTYPE
     SET      technology = 'CMOS'
     WHERE    type_name = 'cell2'

(3)  UPDATE   CELL_PROTOTYPE
     SET      technology = 'CMOS'
     WHERE    type_name = 'cell3'
```

P2 : "Change the functions of gates."

```
(4)  UPDATE   CELL_PROTOTYPE
     SET      function = 'nor-gate'
     WHERE    function = 'gate'
     AND      technology = 'CMOS'

(5)  UPDATE   CELL_PROTOTYPE
     SET      function = 'or-gate'
     WHERE    function = 'gate'
     AND      technology = 'ECL'
```

P3 : "Change the size of CELL1 depending on its function."

```
     $ DCL  FC  FIXED BIN (31)

(6)  $ SELECT  function
       INTO    $ FC
       FROM    CELL_PROTOTYPE
       WHERE   type_name = 'cell1'

     IF  $ FC = 'nor-gate'
     THEN      $ UPDATE CELL_PROTOTYPE
               SET      size = 480
               WHERE    type_name = 'cell1'

(7)  ELSE      $ UPDATE CELL_PROTOTYPE
               SET      size = 420
               WHERE    type_name = 'cell1'
```

Fig. 2.51 (continued)

P4 : "Check whether the technologies of all cells that have to do with CELL3 are identical."

```
         $ DCL  TECH    CHAR(4);
         $ DCL  CTECH   CHAR(4).

(8)      $ SELECT   technology
           INTO     $TECH
           FROM     CELL_PROTOTYPE
           WHERE    type_name = 'cell1';

(9)      $ SELECT   technology
           INTO     $ CTECH
           FROM     CELL_PROTOTYPE
           WHERE    type_name = 'cell2';

           IF CTECH = TECH
(10)       THEN        $ SELECT   technology
                         INTO     $ CTECH
                         FROM     CELL_PROTOTYPE
                         WHERE    type_name = 'cell3';

                         IF CTECH = TECH
                         THEN PRINT ('SUCCESS')
                         ELSE PRINT ('FAILURE')
                         ENDIF
            ELSE PRINT ('FAILURE')
            ENDIF;
```

Fig. 2.51. Four programs on the relational database of Figs. 2.21 and 2.22

a)

type_name	completion	released	technology	function	size
cell 1	851117	T	CMOS	nor-gate	560
cell 2	860120	T	CMOS	nor-gate	560
cell 3	860214	F	CMOS	adder	12628

b)

type_name	completion	released	technology	function	size
cell 1	851117	T	CMOS	or-gate	560
cell 2	860120	T	CMOS	or-gate	560
cell 3	860214	F	CMOS	adder	12628

c)

type_name	completion	released	technology	function	size
cell 1	851117	T	CMOS	nor-gate	560
cell 2	860120	T	CMOS	or-gate	560
cell 3	860214	F	CMOS	adder	12628

Fig. 2.52a–c. Interaction patterns between P1 and P2. **(a)** P2 after P1. **(b)** P1 after P2. **(c)** (1)–(4)–(5)–(2)–(3)

Figure 2.52 illustrates the result of three different types of interaction between two update processes (two **writers**). Cases a) and b) demonstrate the effects of serial executions. Even though the two results differ the corresponding processes are not regarded as interacting. Why is this so? If we study the two writers in detail we discover that each of them guarantees the constraint "the gate functions of cells 1 and 2 are identical", whereas only P2 maintains a stronger rule "ECL gates are always nor, CMOS gates are always or". The latter rule, then, is not a global constraint so that we should not be surprised that it is not enforced by the DBMS. Case c), however, is a clear case of interaction with catastrophic consequences: The database has become inconsistent even though each of the processes was consistent (**writer-writer problem with inconsistency**).

a)

type_name	completion	released	technology	function	size
cell 1	851117	T	CMOS	gate	420
cell 2	860120	T	CMOS	gate	560
cell 3	860214	F	CMOS	adder	12628

b)

type_name	completion	released	technology	function	size
cell 1	851117	T	ECL	gate	420
cell 2	860120	T	CMOS	gate	560
cell 3	860214	F	CMOS	adder	12628

Fig. 2.53a,b. Interaction patterns between P1 and P3. **(a)** P3 after P1 or P1 after P3. **(b)** (6)–(1)–(2)–(3)–(7)

Figure 2.53 demonstrates a second kind of interaction between two writers. Serialization case a) shows what the correct result should be. In case b) the technology of cell 1 remained unchanged because P3 retrieved the old tuple, P1 then replaced it, and P3 subsequently replaced it again but on the basis of the old original. In other words, some of the effects of P1 got lost. This type of interaction is called the **writer-writer problem with lost update.**

Figure 2.54 is different in that one writer interacts with a pure retrieval process (**reader**). Clearly, no damage is possible to the database because P1 is the only process changing it, and by definition P1 guarantees consistency. Nevertheless there is interaction: depending on the interaction pattern the response of P4 to the user may variously be SUCCESS or FAILURE. The reason for the latter cases can be found in the fact that P1 causes intermittent inconsistencies, and it is one or more of these inconsistent states that P4 gets to see. The interaction is referred to as the **reader-writer problem.**

After what has been said it should be quite obvious that two readers do not interact at all and may run concurrently without further precaution (**reader-reader situation**).

Pattern	Response
P4 after P1	success
P1 after P4	success
(1) - (8) - (2) - (9) - (3) - (10)	success
(8) - (1) - (9) - (2) - (10) - (3)	success
(1) - (8) - (9) - (10) - (2) - (3)	failure
(8) - (1) - (2) - (3) - (9) - (10)	failure
(1) - (2) - (8) - (9) - (10) - (3)	failure

Fig. 2.54. Interaction patterns between P1 and P4

How can the interaction be circumvented? Clearly, the simplest solution is to arrange all processes in serial order. However, usually very few will interact so that this approach is overly pessimistic. An improved solution would determine which processes do not interact, and execute them concurrently while serializing the others. Unfortunately, interaction is usually not known beforehand. But there is a dynamic variant to this: Processes are always started right after their queries have arrived (provided other resources like processors and disk space are available); once interaction is detected between two processes they are forced into a serial schedule. Such an interaction-free arrangement of a set of processes is called **process synchronization.**

Interaction is detected in the following way. Once a process wishes to access an item in the database it attempts to attach a lock to it; if as yet no other lock has been set to it the process gains access, otherwise it must wait until the other lock has been removed (in order to let two readers work in parallel, usually two types of locks are distinguished: a **share** lock for purely retrieval purposes, and an **exclusive** lock otherwise). Note that this kind of synchronization is still pessimistic in nature: A lock is attached because interaction is expected to occur, with the concomitant overhead for the management of all the locks.

In order to ensure consistency of the database and correctness of response under any circumstance, setting and managing the locks must obey a **two-phase protocol** :

1. Before a data item is accessed for the first time a lock must be attached to it.
2. If the item already has a share lock only a share lock may be further attached to it. If it has an exclusive lock no other lock is accepted.
3. All locks set by a process are removed only at the end of the process, and then altogether.

The name "two-phase protocol" is derived from the fact that there is one process phase where the number of locks held by the process is steadily growing, and a second where it can only decrease (and, in fact, drops to zero in one step).

The interaction problem may be viewed from a different angle. Had we declared the four processes of Fig. 2.51 as transactions it would have been absolutely obvious that interactions during which intermediate results of a transaction become visible to other transactions must cause problems: The transaction is the unit of consistency, and as such it guarantees consistency only before and after it has been executed. We conclude that the transaction can also be viewed as a "unit of synchronization".

In turn, this view raises the opportunity to leave much of the enforcement of the two-phase protocol to the DBMS provided it is in a position to recognize and monitor the individual transactions. Requirement 1 can be met by intercepting all database operations originating from the transaction, except when the transaction wishes to set locks long before it accesses the corresponding item (in this case the transaction remains responsible for locking). Requirement 2 can always be handled by the DBMS, and likewise 3 when the end of transaction is recognized.

```
P1 :  (1)     UPDATE      CELL_PROTOTYPE
              SET         technology = 'CMOS'
              WHERE       type_name = 'cell1'

      (2)     UPDATE      CELL_PROTOTYPE
              SET         technology = 'CMOS'
              WHERE       type_name = 'cell2'

P2 :  (3)     UPDATE      CELL_PROTOTYPE
              SET         function = 'nor-gate'
              WHERE       type_name = 'cell2'

      (4)     UPDATE      CELL_PROTOTYPE
              SET         function = 'nor-gate'
              WHERE       type_name = 'cell1'
```

Fig. 2.55. Two processes that deadlock under the two-phase lock protocol for the following interaction patterns: (1)–(3)–(2); (1)–(3)–(4); (3)–(1)–(2); (3)–(1)–(4)

It is possible that two transactions **deadlock** because each one of them tries to attach a lock to an item that the other one already holds: see Fig. 2.55 for an example. Since neither of the two interlocking transactions can continue its work, one of the two must be terminated by force. And since the terminated transaction presumably leaves the database in an inconsistent state, the state of the items locked by it must be restored to the one prevailing at the beginning of the transaction. In other words, the terminated transaction must not leave any trace of its execution; it is **rolled back**.

We conclude by mentioning that there is a second class of access synchronization: Optimistic methods assume that access conflicts hardly ever occur, so that transactions may access objects without the need for locking them. The database management system records all accesses and undoes a transaction if an access conflict has occurred during its lifetime (slight variants of the approach are also in use).

2.2.5 Database Recovery

Information is a precious resource for an enterprise. Its corruption or loss may inconvenience the user, cause inappropriate actions or, in the worst case, bankrupt the enterprise. On the other hand, some losses are unavoidable over the course of time. Faulty input may go undetected over a long period and may give rise to ensuing inconsistencies. The computer system may crash due to power failure, hardware failure, software errors, or operating mistakes. Fire, water, or physical mishandling may destroy the storage media. A database system must take precautionary measures (and the enterprise must take organizational measures) to recover from situations like these. The steps taken to restore the database to the state it was in before the loss occurred, and the preparatory actions taken during normal operation are the responsibility of a **recovery** subsystem within the DBMS. Its objective is to give to the application world the impression of an inviolable, **persistent** database.

Ideally, the DBMS reconstructs the database to the state it was in right before the abnormal or catastrophic situation occured. In certain cases, however, – power failure is one – some of the information needed is permanently lost so that we have to set a somewhat less ambitious goal: Reconstruct the database to the last consistent state it was in before the failure (even more moderate goals are conceivable, with a concomitant reduction in overhead; we shall not further consider them here).

As we discussed before, the concept of consistency is tightly associated with the concept of transaction. It is small wonder, then, that we consider a transaction not only as the "unit of consistency" but also as the "unit of recovery". Rephrasing our objective: On recovery the DBMS must restore the effects of all transactions completed up to the failure point, and remove any traces of all transactions that were in the midst of execution when the failure occured.

In order to state our expectations more precisely we should examine the three failure situations in more detail.

1. **Transaction failure.** A transaction cannot be brought to completion because it aborts itself, e.g. , when the user decides that he cannot complete a design for want of a particular cell prototype, or because it is aborted from outside, e.g., when it must be rolled back to resolve a deadlock. This failure situation is characterized by intact main and peripheral storage, and a single affected transaction.
2. **System failure.** The hardware, operating system, or DBMS is abruptly brought to a halt in a state from which it cannot resume execution. As a consequence, all executing transactions are automatically aborted. The situation differs from the previous one in that the contents of some or all of main storage are lost while peripheral storage remains intact; also, an arbitrary number of transactions are affected.
3. **Storage media failure.** Besides by violent destruction, storage media - or the information they hold - may become lost due to storage device failure (the infamous head crash on disks is an example) or environmental failures

(high temperatures or chemical reactions because of faulty air conditioning are examples here). The situation is described by the loss of some of the permanently stored data; nothing can be said about main storage.

A prime requisite for recovery from all three situations is the ability to restore a transaction to one of the states "not executed" or "completed". The principal solution is fairly straightforward: All changes to the database during execution are considered temporary and are only made permanent ("commited to the database") at the end of transaction (the action is, therefore, called a **transaction commit**). As a user one should be interested in just this behavior, not in how it is technically achieved. Nevertheless, here are a few examples of technical solutions in order to give the reader a feeling for the price in performance to be paid for the recovery property.

a) Keep all changes in main memory (provided it is large enough) and write them to peripheral storage during commit.
b) Write all changes to a different section of peripheral storage and switch to this section at commit.
c) Save the old state of the data to be changed on a separate section of peripheral storage (**log file**; the old data are called the **before-image** of the transaction), then write the new data to the old place on peripheral storage.
d) Keep all changes in main memory and write them to peripheral storage some time after commit.

Figure 2.56 illustrates the actions necessary for transaction failure. Rollback on transaction failure is trivially achieved in cases a), b) and d) by inaction, and in case c) by retransmitting the before-image back to its old place.

Case d) is special in that, unlike the other cases, it is vulnerable to system failure: From the user's viewpoint the transaction changes are permanent, but technically the changes have been lost if failure occured before writing them to peripheral storage. The solution is to save the new data (called the **after-image** of the transaction) on the log file (this can usually be done faster than writing them to their permanent place). Consequently, as Fig. 2.57 illustrates, case d) differs in its actions for system failure from Fig. 2.56, whereas the others remain the same.

Media failure is more complicated to handle because of the loss of permanent data. Figure 2.58 illustrates a typical procedure. At suitably chosen intervals a **database dump** (usually on magnetic tape) is taken of the entire database while it is in a consistent state, i.e., while no writer transactions are executing. Between intervals, a log is kept in the log file of all transactions taking place. One specific way, though not the only one, in which to do this is to record the after-image of each transaction. Figure 2.59 illustrates the actions necessary in this situation.

Take as an example for P1 to P4 in Fig. 2.58 the four transactions of Fig. 2.51, suitably arranged over time to keep them from interacting. Suppose now a catastrophic event as indicated in the figure. Then the database (or the lost part of it) is reconstructed in two steps:

Transaction	Recovery preparation case a)	case b)	case c)	case d)
BEGIN_TRANS				
UPDATE CELL_PROTOTYPE SET function = 'nor-gate' WHERE function = 'gate' AND technology = 'CMOS'	no action	no action	no action	no action
UPDATE CELL_PROTOTYPE SET function = 'or-gate' WHERE function = 'gate' AND technology = 'ECL'	modified tuples in main memory	write modified tuples to a new section of peripheral storage	save old tuples on log file; replace old tuples by modified ones on peripheral storage	modified tuples in main memory
END_TRANS	commit: write modified tuples from main memory to peripheral storage	commit: make new section the current one	commit: remove before-image from log file	commit: no action (system writes modified tuples to peripheral storage at a later time)

Fig. 2.56. Actions to be taken during transaction execution in order to facilitate eventual recovery from transaction failure illustrated for transaction P2 from Fig. 2.51 and database state as in Fig. 2.22. For a description of cases a), b), c) and d) see the text

Transaction	Recovery preparation: case a)	case b)	case c)	case d)
BEGIN_TRANS				
UPDATE CELL_PROTOTYPE SET function = 'nor-gate' WHERE function = 'gate' AND technology = 'CMOS'	no action	no action	no action	no action
UPDATE CELL_PROTOTYPE SET function = 'or-gate' WHERE function = 'gate' AND technology = 'ECL'	modified tuples in main memory	write modified tuples to a new section of peripheral storage	save old tuples on log file; replace old tuples by modified ones on peripheral storage	modified tuples in main memory
END_TRANS	commit: write modified tuples from main memory to peripheral storage	commit: make new section the current one	commit: remove before-image from log file	commit: write modified tuples from main memory to log file; after system wrote from main memory to peripheral storage: remove after-image from log file

Fig. 2.57. Modification of Fig. 2.56 for recovery from system failure

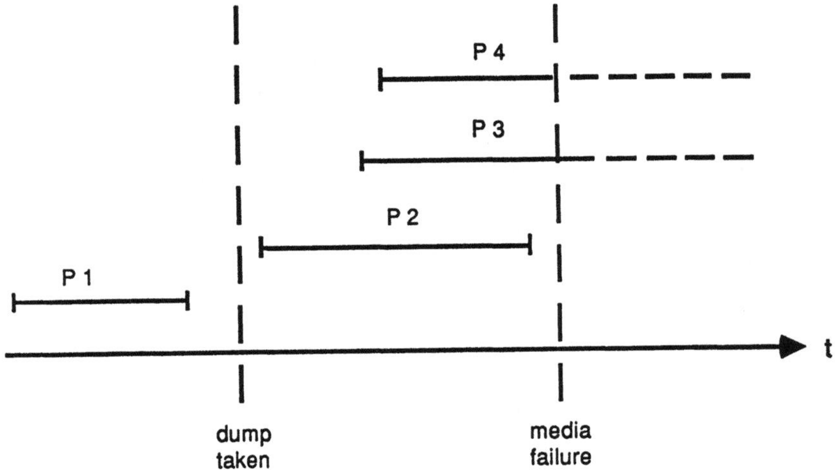

Fig. 2.58. Transaction history with four transactions P1, P2, P3, P4, and a dump. Media failure occurs during execution of P3 and P4. The solid bars indicate the time span during whitch the transactions are executing; they are not necessarily active over the entire time due to suspension to keep them from interacting

1. **Restore** the database state at dump time by copying the dump back to a presumably new medium. Note that this automatically restores the effect of transaction P1.
2. **Redo** those transactions that started after dump time and completed before media failure. In this case P2 must be redone, by copying its after-image to the new medium.

If several transactions must be redone, their serial order is important. Hence, this order must be recorded as part of the log file (Fig. 2.59).

Evidently it makes sense to combine the actions for all three failure situations into one log file. We could then also deal with the situation of an inconsistent dump. Suppose we take a dump while all three, P2, P3, and P4, are running. Then the steps look slightly different:

1. **Restore** the database state at dump time.
2. **Redo** all transactions that completed after dump time but before media failure, by copying their after-images. In Fig. 2.58 this would again be P2.
3. **Undo** all transactions that started before dump time and did not complete before media failure, in case c) by copying their before-images. In Fig. 2.58 P3 and P4 would be affected, although no action would actually take place for P4 because it is a pure reader.

Note that to be able to recover depends critically on the inviolability of the dump tapes and the log file.

Transaction	Recovery preparation			
	case a)	case b)	case c)	case d)
BEGIN_TRANS	note transaction on log file	note transaction on log file	note transaction on log file	note transaction on log file
UPDATE CELL_PROTOTYPE SET function = 'nor-gate' WHERE function = 'gate' AND technology = 'CMOS'	no action	no action	no action	no action
UPDATE CELL_PROTOTYPE SET function = 'or-gate' WHERE function = 'gate' AND technology = 'ECL'	modified tuples in main memory	write modified tuples to a new section of peripheral storage and to log file	save old tuples on log file; replace old tuples by modified ones on peripheral storage and write modified ones to log file	modified tuples in main memory
END_TRANS	commit: write modified tuples from main memory to peripheral storage and log file; mark transaction completed	commit: make new section the current one; mark transaction completed	commit: mark transaction completed	commit: write modified tuples from main memory to log file; mark transaction as completed (do not remove after-image!)

Fig. 2.59. Modification of Fig. 2.56 for recovery from media failure

2.2.6 Transaction Management

The notion of transaction is central to the dynamics of a database: The transaction is the unit of consistency and, as a consequence, the unit of synchronization and the unit of recovery. Or, to phrase it differently, a transaction is characterized by the following properties:

- **Consistency:** A transaction, when applied to a consistent database state, results in a consistent database state.
- **Atomicity:** Once it has been started, a transaction is either completed or it leaves no changes in the database.
- **Persistency:** The result of a completed transaction remains in the database forever unless it is explicitly changed by a subsequent transaction.
- **Separability:** Database states referred to during the execution of a transaction cannot be shared by a second transaction (this property, essentially a consequence of atomicity, is often included there).

Which sequence of database operations forms a transaction depends on what one considers to be consistent states and transitions of the database. The narrowest interpretation is one that just refers to the explicitly stated consistency constraints; transactions from this viewpoint are often called **database transactions.** The widest interpretation considers entire units of work in the application environment, where intermediate stages of the work are deemed incomplete, contradictory, or prone to resumption at earlier points, revisions, or cancellations. Typical transactions in that sense (**application transactions**) are reservations (airlines, car rentals, hotels), bookings (vacation trips), retail sales (supermarket checkouts), money transfers, and planning and filling of orders, to name just a few. In the same sense the design of a complete cell prototype may be viewed as a transaction, too. The former examples – representatives of what one may call **commercial transactions** – are usually of the duration of seconds or minutes, and affect only very few items in a database, whereas the latter, an **engineering transaction**, could last over weeks and may during its course relate to large amounts of data. The present state of the art is predominantly oriented towards commercial transactions.

Application transactions may contain one or more database transactions. From a user's point of view, however, the distinction remains irrelevant: all that he initiates are application transactions and, indeed, he expects these to have the properties mentioned above.

The implementation techniques for consistency enforcement, synchronization of multiple transactions, and recovery from transaction, system, and media failures are highly interdependent so that they are concentrated within a single **transaction management** subsystem. In fact, however, the notion of application transaction is neither new nor confined to database management systems, as the examples illustrate. For a long time vendors have offered **data communication systems** as enhancements to operating systems, that specifically support interactive applications. They assume that any of a number of preplanned ("canned") transaction programs are started at unknown and in-

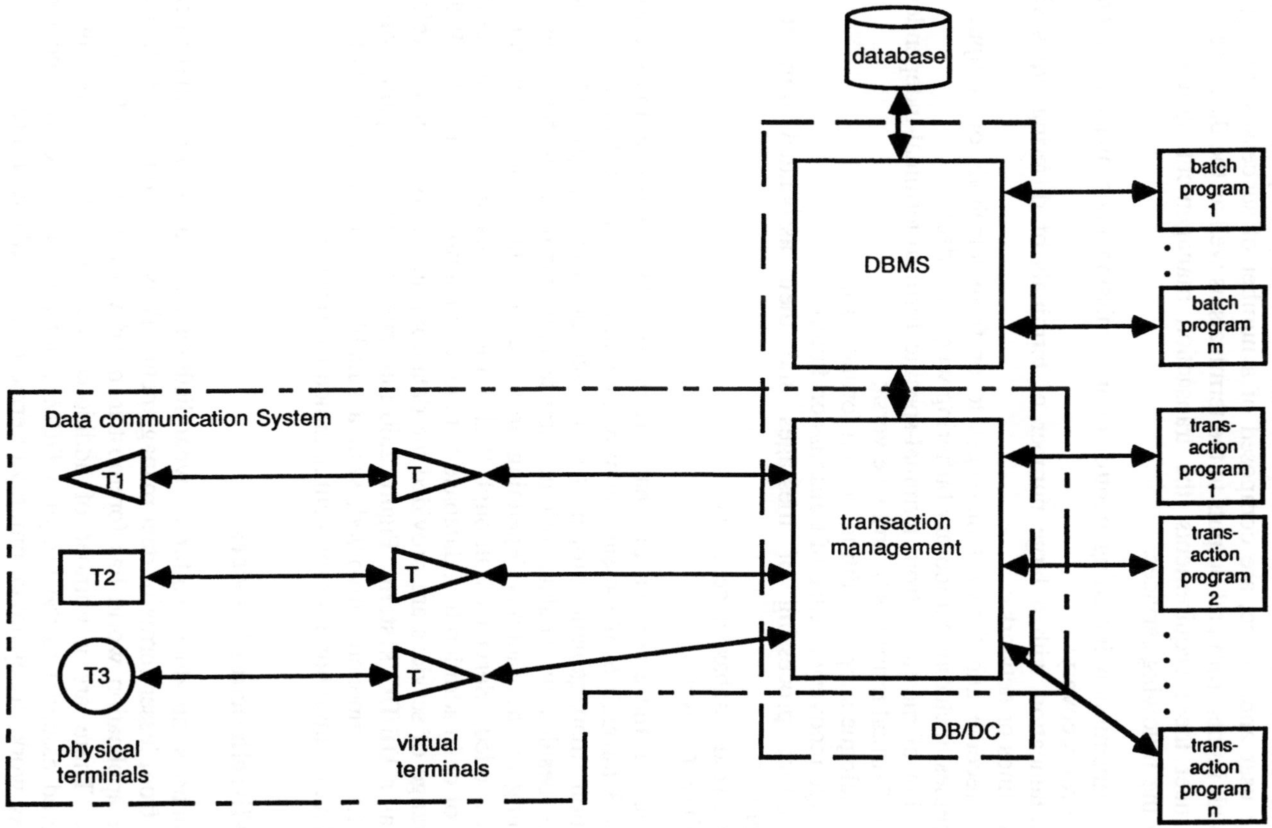

Fig. 2.60. Basic organization of a DB/DC system

dependent times via messages from a wide variety of terminals and executed concurrently. The same program may be activated repeatedly and concurrently with the same or different sets of parameters. Usually the programs interact with the user and, hence, are composed of a number of successive dialogue steps, with inputs from and outputs to the terminals as well as the database.

To meet these requirements, the transaction management system must provide the following services.

1. data transmission including control of transmission lines and equipment, and error recovery;
2. communication with a large number of terminals of differing types via direct lines or networks;
3. send, receive and forward messages to or from terminals or computing processes, including temporary buffering where needed;
4. mapping of messages from terminal-specific form to terminal-independent form ("virtual terminals"), and vice versa;
5. user dialogue support (editing, screen formatting);
6. message recovery in case of transmission failure;
7. concurrent processing of messages and their associated transaction programs;
8. centralized database access; and
9. database recovery.

Services 1 and 2 are part of what is referred to as network administration, services 3 through 6 of message administration, and services 7 through 9 of transaction management proper (or transaction monitoring). The latter includes, besides the database-oriented transaction management, functions for scheduling the transactions depending on the availability of all computer resources needed. Services 7 through 9 also apply if files are to be processed instead of or in addition to a database but are much simpler in this case. If the entire range of services are provided the data communication system is referred to as a **DB/DC system**. Figure 2.60 illustrates the basic organization of the system components involved, with a suitable division of labor between DBMS and transaction management assumed for services 7, 8, and 9.

2.2.7 Miscellaneous Services

The functions introduced so far are what distinguishes a database management system from lesser forms of data management. They represent those features that the application world has foremost in mind when it decides to install a DBMS. There are a number of additional facilities that are not always considered essential to the use of a DBMS, but add to the conveniences of everyday work. They are commonly referred to as **database utilities**.

2.2.7.1 Protection

Database security is concerned with measures that guarantee that confidential data is available only to authorized persons, so that privacy requirements are satisfied and enterprise secrets guarded; and that the database accurately reflects the state of the enterprise, i.e., is protected against either malicious or accidental destruction or corruption. Sections 2.2.4, 2.2.5 and 2.2.6 dealt with those security aspects that are DBMS immanent. **Database protection** is concerned with all other security aspects. In general, protection is seen in the context of computer misuse such as illegal inspection of proprietary information, e.g., for gaining an advantage in negotiations, for blackmail, or for espionage; corruption or destruction of data for sabotage, e.g., disruption of technical processes, production schedules, or financial transactions. In addition, protection should be seen as a way to guard the database against accidental actions that lead to database corruption, such as destroying a vital relation just by mistyping a relation name, and that are expensive to recover from.

Database protection is based on the principle of **authorized access.** Authorization is the specification of rules about who has what type of access to what information. An authorized access abides by these **access rules.** The DBMS component enforcing the rules is the **access control** subsystem. One may visualize the totality of access rules arranged in an **access matrix,** with the rows identifying the "subjects" (system users, but also application programs), the columns identifying the "objects" (the database items, but also programs that access or manipulate the database), and the matrix elements specifying the actions permitted on objects by the subjects. See Fig. 2.61 for an example. Specifying and updating the matrix is an ongoing process

Objects

Subjects	CELL_PROTOTYPE	PORT	CELL_INSTANCE	ADJOINMENT	RECTANGLE	CONNECTION	CELL
A. Smith	I, U	S, I, U, D	S, I, U, D	S, I, U, D	S, I, U, D	S, I, U, D	--
J. Doe	S, D	S	S	S	S	S	S
B. Miller	--	S	--	--	--	--	S
Logic-analyzer	S	S	S	S	S	S	--
Layout-generator	S, I	S	S	S	S	S	--

Fig. 2.61. An access matrix for the database of Fig. 2.21. Protection granularity "is" relations. Note that one of the objects is a view (Fig. 2.39). The subjects are either persons or application programs. Access operations are: S: select, I: insert, D: delete, U: update

relegated to an especially trustworthy security administrator who is provided with specific DDL elements (PRIVACY LOCKS in the network DDL, a GRANT operation in SQL) or with a separate authorization language. Needless to say, the access matrix (or its implemented counterpart) must be protected especially carefully.

During DBMS operation, protection involves two distinct tasks. **Authentication** determines whether the user is the subject he claims to be, and is based on personal traits such as passwords, code cards, finger and voice prints, and handshaking procedures, to name a few. **Access control** makes sure that an authenticated subject performs only those actions on a given object that conform to the access matrix.

More advanced protection facilities provide the means for context- or content-dependent access rules, for the inclusion of views, and for data encryption.

2.2.7.2 Mass Data Input and Output

Data are often generated outside a DBMS, or transferred from a file or some other DBMS, or moved to secondary peripheral storage because they are only rarely used. Entering or removing these data item by item via DML operations is a rather cumbersome and slow process. Instead, the DBMS provides **load/unload** services to perform mass data input and output in a more efficient manner, both as far as processing time and space utilization are concerned.

In particular, the unload service is useful for **archiving** those parts of the database that are only needed on rare occasions, e.g., for auditing purposes, or to maintain the history of the evolution of a design, or to keep legal records over a time period as specified by law as in case of health records or examination records. To include the archival data in the database proper would burden its performance and complicate data protection. Note that transaction logs collected for recovery purposes may also sometimes become part of an archive.

2.2.7.3 Data Dictionary

The precise semantics of a DML operation is only determined on execution, by inspecting and interpreting the database schema. Furthermore, as will be discussed in more detail in Sect. 2.2.9, even the algorithmic solution for implementing the operation can be influenced during execution by controlling the implementation details via a storage (or physical) schema (the internal schema of Fig. 2.1). In turn, the schemas are created, modified, updated and inspected by a special group of persons, the **database administration**. One could view these schemas as a database in its own right, which is logically distinct from the database containing the raw data; to distinguish it from the latter it is often referred to as a **meta-database**. Users of the meta-database are, besides the database administration and the DBMS managing the raw data, other personnel like managers and system analysts. In fact, once the separation exists, one may include in the meta-database other information such as a list of transaction programs together with the subschemas they use, cross-reference lists, system-analytical results like semantic schemas or usage statistics. Facilities of this kind if, from a user's point of view, they appear fully separated from the DBMS, are called **data dictionaries**.

2.2.8 System Organization and Environment

2.2.8.1 Distribution Aspects

A database management system depends on an environment that furnishes the technical resources for supporting it, and the human resources for utilizing it. Turning first to the latter aspect, the discussion on transaction management (Fig. 2.60) already gives the flavor of how a DBMS interacts with the user world. Generally speaking, since DBMS are a means for integration, a DBMS must give all users of an enterprise the impression of one single, centralized

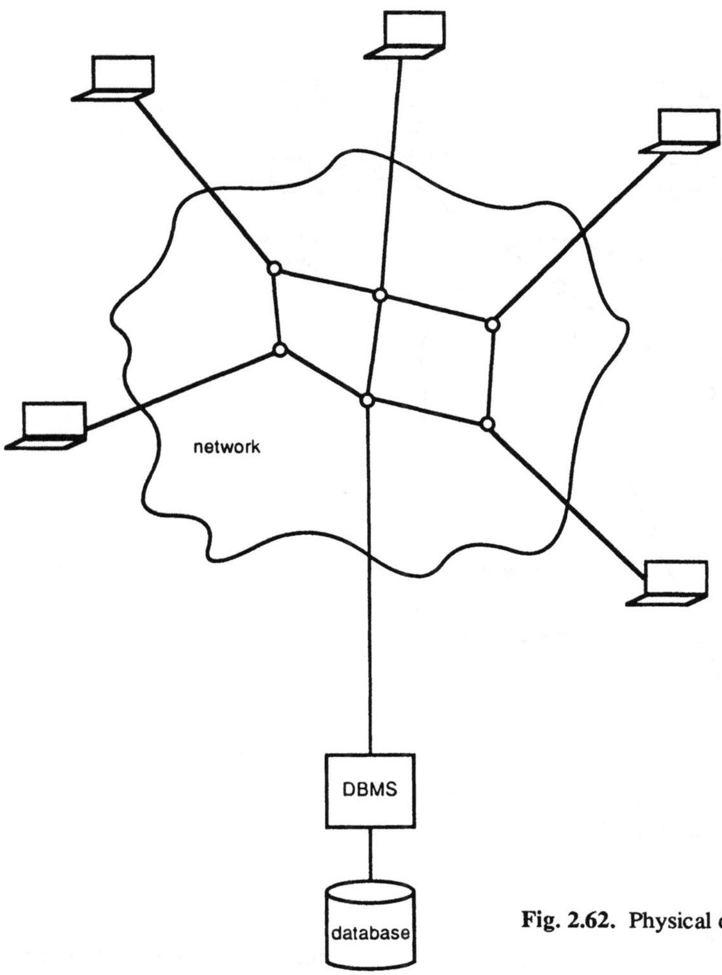

Fig. 2.62. Physical database centralization

database. This does not imply, however, that the data or the management functions are physically located at a single site: instead they may be distributed over several geographical sites.
Basically, we distinguish three ways of organizing a DBMS and its database.

- **Physical centralization.** A central and usually large computing facility is provided that is accessible from numerous terminals via a local or wide area network (Fig. 2.62). An advantage is that proportionately little overhead is involved in central control including transaction scheduling, consistency enforcement, synchronization and recovery. A heavy price is paid, however, in terms of overall systems reliability (if the central facility breaks down the entire database becomes inaccessible), growth potential (performance bottlenecks can only be overcome by extending the central facility in rather crude steps), and transmission costs.
- **Distributed database.** A number of sites with their own – usually smaller – computing facilities, each with a complete copy of the DBMS but with its own local database, are interconnected via a network (Fig. 2.63). This

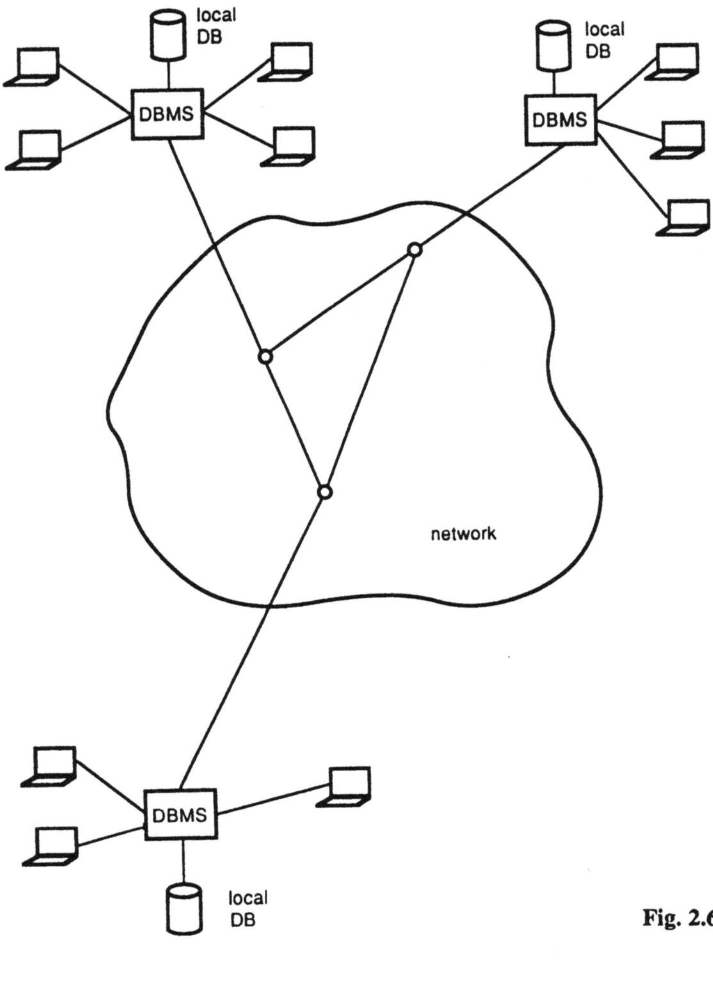

Fig. 2.63. Distributed database

type of organization overcomes some of the weaknesses of the centralized approach: Transmission costs are low provided much of the database access is to local data rather than to those at other sites; the reliability is usually much higher because disruptions in the network still permit local processing, and the breakdown of one site affects other sites only to the extent that they need access to its data; incremental growth is easily achieved by adding a new site or extending a relatively small computer at one site. Unfortunately, there is also a price to pay for this: The overhead for coordinating the activities at several sites so that the transaction properties remain satisfied is much higher than in the centralized case and may even wipe out many of the other gains.

- **Server/client organization.** This approach looks somewhat like a compromise betweeen the first two approaches. A central site maintains the database, providing services such as bulk data access, protection, and limited forms of consistency enforcement, synchronization, and recovery. The transaction programs are executed at local sites to which the required data are shipped at transaction begin from the central site via network, and they

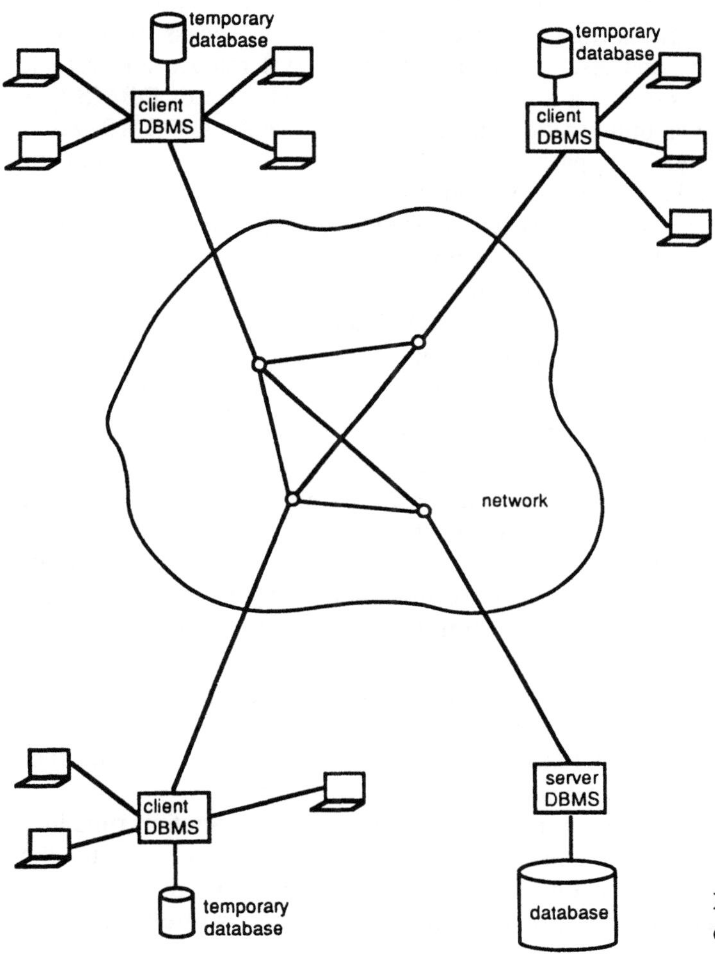

Fig. 2.64. Server/client database organization

are returned to the latter at transaction end (Fig. 2.64). In order to manipulate the data during the transaction the local sites have their own DBMS and transaction management, again in a functionally limited form, e.g., without synchronization. Consequently, because of the division of labor between central (server) and local (client) sites, with only partial overlap of the functions needed, both the central and the local DBMS can be simplified versions of a DBMS in the central approach. This type of organization is currently under active research.

As far as the data definition and data manipulation functions are concerned, a user should remain unaware of the type of DBMS organization. It is important only to the system analyst and the database administrator.

2.2.8.2 Hardware and Operating System Aspects

Turning now to the issue of technical resources, those that have the biggest effect on the performance of a DBMS other than the DBMS implementation itself are the peripheral storage media used, and the operating system supporting it. Figure 2.65 gives an overview of the present technology for peripheral storage, with storage capacity, storage cost, access time, and writeability the decisive factors. As far as operating systems are concerned, even the more modern systems are not optimally disposed towards DBMS and often tend to dilute the optimization measures taken within a DBMS. As a consequence, most DBMS bypass many of the high-level services that an operating system provides, and re-implement some of these in accordance with their own needs on the basis of lower-level services. Unfortunately, this makes a DBMS especially vulnerable to revisions in the underlying operating system. Other services such as virtual memory management cannot be bypassed at all although they are detrimental to DBMS performance.

Since the issues involved are highly technical we shall not go into details here, except for two aspects that may at times become visible to the user.

- **File management.** Ultimately, the database consisting of raw data, database schema, and auxiliary structures to support associative access (**access paths**, e.g., indexes) must be mapped to a number of files. When a DBMS is installed, and a database is created, the files needed must be defined including numerous parameters concerning file access method, storage space organization, record formats and lengths, and buffering techniques, to name just a few. An unfortunate choice of parameters may have a negative impact on performance.
- **Task management.** Today's operating systems are based on the principle of concurrent execution of processes. Consequently, a straightforward organization of transactions would associate a process with each transaction. Still, there are several alternatives.
  a) **Process-per-transaction model.** Each concurrent transaction runs in a separate user process (task). The DBMS is "linked in" to the transaction program within the task. Scheduling is done by the operating system according to its own strategy which is rarely geared to

medium	technology	no. of tracks	useable capacity (MByte)	write-ability	removability	volatility	av.direct access time (msec)
<u>mainframes and minicomputers</u>							
high-capacity removable magnetic disk	magnetic	16.000	400	unlimited	yes	head crash	24
high-capacity fixed magnetic disk	magnetic		2.700	unlimited	no	head crash	20 .. 30
<u>workstations and personal computers</u>							
1/4" magnetic tape (streamer)	magnetic	36	90	unlimited	yes	dust, climate, wear, mechanical damage	-
5 1/4" floppy disk	magnetic	96	3	unlimited	yes	dust, climate, wear	90
5 1/4" winchester disk	magnetic	2.000	300	unlimited	no	head crash	25..30
optical disk (5 1/4" .. 12")	optical	15.000	240. .2.000	once	yes	non-volatile	60
compact disk	optical	30.000	4.000	no (data are pur-chased)	yes	non-volatile	60

Fig. 2.65. Technical characteristics (peak rates) of typical peripheral storage media

the specific requirements of database usage. Each task has its own copy of the DBMS code unless it is re-entrant. Because some interaction between the task DBMS is necessary to guarantee the transaction properties, several shared data structures must be managed, and access to them synchronized by the operating system. These data structures tend to develop into serious performance bottlenecks.

b) **Independent DBMS process.** To overcome the difficulties, one may organize the DBMS into a separate, permanently executing task. Communication between the user tasks that now include just the transaction program, and the DBMS task is by messages. A real advantage is gained only if the scheduling of the transactions can be made the responsibility of the DBMS task rather than being left to the operating system. Even then, the high number of task switches undercut performance.

Which strategy one should choose – if there is any choice at all – can only be decided after a careful analysis of all technical issues involved. This clearly is the responsibility of the database administrator, not of the application user.

2.2.9 Performance Control

The external factors that affect the performance of a DBMS were just discussed. There are also a variety of DBMS-internal factors that have an influence. Many DBMS offer some limited means for controlling these. One may interpret them as parameters that permit a more precise configuration of the DBMS or the detailed storage organization of the database. To give the flavor, we turn our attention to Fig. 2.66 which shows the top-to-bottom functional organization of a DBMS into a number of successive levels. These reflect the stepwise implementation of the data structures satisfying the schema, and of the DML operators; transaction management issues have been omitted.

The topmost interface is considered to be set-oriented such as for the relational data model, and the one below to be navigational and record-oriented as for the hierarchical and network data models. What the architecture suggests is that a set-oriented interface is implemented in terms of single-record accesses, e.g., a relational interface in terms of tuple-wise access. Although this is not absolutely necessary (the implementation may be based directly on the internal record interface), it has the advantage of defining set-like structures that directly support expensive operations like relational joins. The rich semantics of a navigational interface (see Sects. 2.2.1.2 and 2.2.2.3) are mapped to a more simple-minded internal record interface. What that interface allows in particular is the selection of specific techniques in order to implement the access paths defined on the higher levels, such as unique or duplicate keys for direct associative access, sequential access within relations, segment classes or sets, or edges for navigation from one record to a new one. Typical selections involve overall implementation choices between index and

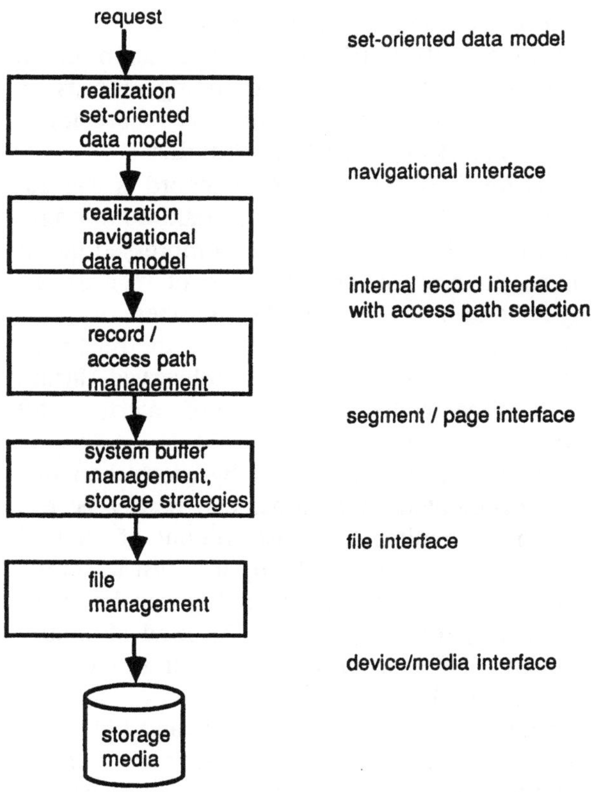

Fig. 2.66. Multi-level DBMS architecture

hashing techniques, and within these the choices of parameters. Both the internal records and the access paths are implemented in terms of storage pages that are themselves grouped into segments. In turn, these are mapped to storage blocks and files. The differentiation between segments and files is necessary because of a buffer management that is tailored to the needs of the DBMS, highly specific storage strategies, and the particular attention paid to transaction requirements.

What kinds of control may a database administrator exercise within such a database system?

1. **Declaration of associative, sequential and navigational access paths.** This is usually done on a level where the logical structure is fully preserved, e.g., on the set-oriented level for the relational model (witness the declaration of indexes, Fig. 2.38) and on the navigational level for the network model (note the declaration of keys and sets, Fig. 2.34).
2. **Implementation of access paths.** For keys the choice is between index and hashing techniques, for sequential and navigational access paths one may decide between various pointer techniques or physical arrangements. Control is exercised on the internal record level.

3. **Physical clustering.** If one knows beforehand that a certain collection of data is usually manipulated in its entirety (e.g., a complete relation, or all members of a set instance), one may place the corresponding records on peripheral storage in physical proximity. Again, the internal record interface, or the segment interface instead, is the place to control this.
4. **Storage space organization.** If much of the space of a record is wasted, e.g., because record fields hold much less than their maximum-length strings, or field values are undefined, the record may be compressed into a smaller unit. On the other hand, if insertion or expansion of records are frequent within storage pages, one may reduce the number of page overflows and the overhead associated with it by controlling the fill factor for the pages within a segment. Compression must be observed during mapping of the navigational interface to the internal record interface, fill factors can be influenced on the segment level.
5. **Query optimization.** A given problem can be stated by more than one relational query. A user of a set-oriented data model can hardly be expected to determine the one which requires the least amount of internal processing. Rather this should be done automatically by the DBMS on the basis of the original query submitted. However, optimality can be influenced by defining indexes specifically for that pupose. This kind of control is exercised on the set-oriented level. This level may also address the issue of precompilation of frequently repeated queries.
6. **Degree of consistency.** The degree of concurrency can be raised if one is willing to relax the strict consistency requirements for transactions, e.g., by permitting reader-writer interactions while still preserving the consistency of the database. Specifying the desired degree is a matter for the segment level.

In earlier systems the corresponding control statements were placed within suitable DDL statements of a database schema. The modern interpretation of a database schema as one that exclusively reflects the functional aspects of the application precludes this approach. Consequently, in modern systems the technical factors and parameters are relegated to a **storage schema** expressed by the database administrator using a **storage structure language (SSL).**

2.2.10 Further Reading

The following books give a more comprehensive introduction, illustrated by numerous examples, of database issues. They require only a general background in computer science.

Date C.J: *An Introduction to Database Systems.* 3rd ed., 2 volumes. Addison-Wesley Publ. Co. 1981

Fernandez E.B., Summers R.C., Wood C.: *Database Security and Integrity.* Addison-Wesley Publ. Co. 1981

Korth H.F., Silberschatz A.: *Database System Concepts.* McGraw-Hill Book Co., 1986

Martin J.: *Principles of Data-Base Management.* Prentice-Hall, Inc. 1976

Tsichritzis D.C., Lochovsky F.H.: *Data Models.* Prentice-Hall, Inc. 1982

More detailed discussions of technical issues are found in

Cardenas A.F.: *Data Base Management Systems.* Allyn and Bacon, Inc. 1979
Martin J.: *Computer Data-Base Organization.* Prentice-Hall, Inc. 1975
Wiederhold G.: *Database Design.* 2nd. ed. McGraw-Hill Book Co., 1983

The theoretical background of database technology is presented in

Ullman J.D.: *Principles of Database Systems.* 2nd ed. Pitman Publ. Ltd., 1982

2.3 Database Systems for Engineering Applications: A New Focus

2.3.1 Perspectives

Though database management systems have become an indispensible component of administrative and business application systems, their usefulness to engineering applications is still a matter of dispute. The reason usually given is the poor performance that database management systems show when they are used within design tasks such as CAD, in mechanical engineering, VLSI design, and software development. Lack of performance, then, is the most evident shortcoming of commercially available database technology with respect to engineering applications. Nevertheless, it is not the only one and one may argue that poor performance itself is (at least in part) merely the result of several other deficiencies.

Database management systems have been designed for special classes of applications, whose characteristics are relatively simple ("flat") data structures, logical units of work (transactions) of short duration, with few records being involved in one transaction. One cannot realistically expect that database management systems do their job equally well when applied to other application areas which do not share these characteristics at all.

We might conclude with these remarks, were it not for the fact that the shortcomings of current database technology for engineering applications have already been widely recognized, and ongoing research at numerous places is actively directed towards overcoming them by developing new and more suitable database techniques. Since first results will be available in the not-too-distant future, a forward-looking approach in the planning of engineering databases should take the developments into account. While these are the subject of Sect. 2.4, the present section discusses in some detail how the requirements of engineering databases differ from those of the more traditional applications. This discussion will be structured according to the organization of Sect. 2.2. This should provide the reader with a more thorough motivation for and understanding of the techniques introduced in Sect. 2.4.

2.3.2 Data Model

Our discussion in Sect. 2.2 already uncovered several shortcomings in existing data models. Foremost among them is that the relationship concept lacks expressive power, be it the set concept which forces one to resort to link records all too often, destroying much of the semantics that way; or be it the relation concept where order cannot be imposed, and functional interdependencies require unnatural constructs such as alternate keys and referential integrity that are offered by very few systems. None of them allows the placement of lower or upper limits on the number of relationships a data item may enter into (remember the fan-out factor in design rule 3!).

Treatment of recursiveness is haphazard at best, while a systematic approach is needed. For example, cells may adjoin other cells, or in a parts list parts may be composed of other parts. Both situations cannot be expressed in the network model, or at least not in the purist version. The relational model does not prohibit such recursion in relations, but then, unfortunately, the DML cannot deal with recursion. On the other hand, if one wishes to prohibit accidental recursion (to satisfy, e.g., design rule 8), one must resort all too soon to consistency constraints.

Unsatisfactory treatment of recursiveness is a symptom of a deeper problem. Design may have to be viewed at different times on different levels of abstraction. For example, a cell may either be treated in its totality (e.g., when it is to be shipped from the server to a client workstation, or to be deleted), or in its structural composition. The latter view is supported reasonably well by a navigational environment. The former view, however, forces the system into a lengthy and expensive process of collecting the components via navigation in the network model, or of joining numerous relations in the relational model (try to write down a SELECT query to retrieve the entire topology of CELL3!). Systems that allow collective views on different levels of abstraction are called **object-oriented** (where **object** refers to an ensemble that a user wants to deal with as a whole). The classical data models lack object-orientation.

Another aspect of abstraction is what is commonly referred to as **generalization**. Generalization abstracts from some of the details of data items. For example, a cell prototype can be considered a generalization of cell instances, abstracting from the positional information. In turn, the cell instances inherit certain properties from their prototypes, such as the technology. None of the classical data models accounts for such inheritance semantics.

The cell of Fig. 2.3 is the result of a lengthy design process. A careful analysis of design processes turns up a number of phenomena and artifacts that a data model for engineering applications should take into account. In general, development processes are iterative or trial-and-error activities. Usually this iteration is performed in an organized manner following some kind of a life cycle or phase plan. Each phase has its output which is useful in its own right, either because it becomes a part of the overall result of the design process, or because it is an intermediate result that forms the basis for iterations. To be a little more explicit, from the way they are processed we can distinguish several kinds of phase results:

1. Each phase of a design process yields results which describe the design object from a phase-specific point of view. In our example of VLSI design, such descriptions are functional descriptions, geometrical/physical layouts, and manufacturing information, to name only a few. As these results contribute a particular aspect to the entire design they are called *representations.*
2. The result of a design step, even though it may have been used as input to a next step, should not be deleted. The designer may wish to return to it because it was found in later design steps to contain a flaw, or the designer may improve on it, or he may extend its functionality. In fact, usually the designer wishes to keep a history of the design result as it evolves over time. Elements of such a history log are referred to as *revisions.*
3. Given the output of one step, a designer may process it in more than one way, each one legitimate, although subject to different constraints. Consider the functional description of a VLSI circuit that may be processed into several geometrical layouts. Such coexisting results formed from the same basis are called *alternatives.*

Representations, revisions and alternatives may be considered different ways in which one object can exist in multiple instances, or in multiple *versions.* When occurring together, one may organize them into a hierarchy of versions, as in Fig. 2.67 for the VLSI example.

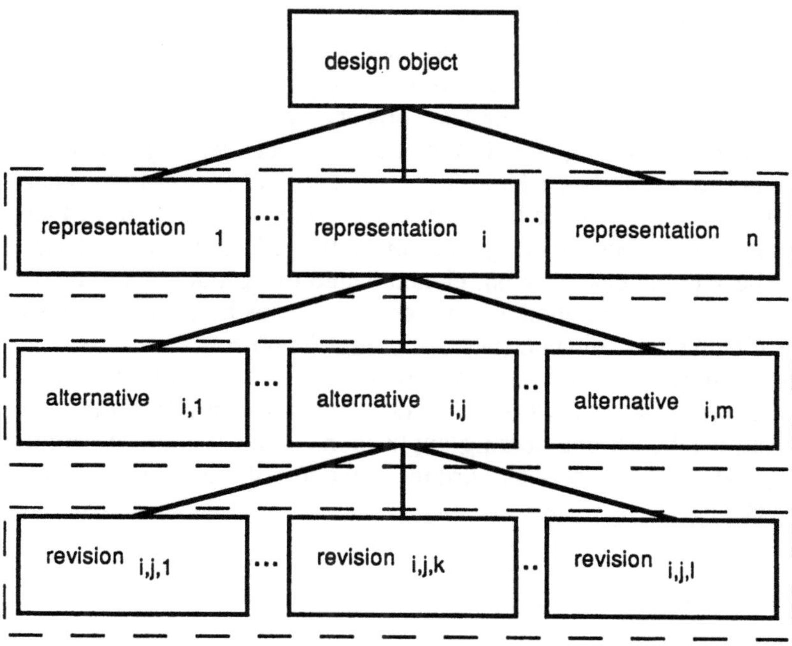

Fig. 2.67. Version hierarchy of a VLSI design object

Whereas revisions reflect the organizational structure of the design object, there is also a technically motivated structure to it that has to do with the design strategies of *top-down* design (where the designer starts with an object and then decomposes it into its subobjects) and *bottom-up* design (where the designer forms one complex object by composing it from several subobjects). The result is the object-orientation mentioned above. As an example consider our VLSI circuit. Obviously, an object of type 'CELL3' consists of several subobjects of type 'CELL1' and 'CELL2' which are connected by several paths. As the subobjects might themselves consist of subobjects (and so on) we talk about *object hierarchies* or *complex objects*.

A suitable data model must support the characteristic properties of both design processes and design objects. To put things together, a data model suitable for engineering applications should meet the following requirements:

DM1 **Complex object support.** The data model must be *object-oriented*, i.e., allow modeling and processing of complexly structured design objects as logical units. If the user wishes to make the structure known to the database management system so that it can control the consistency of the structure, and also process it efficiently, the data model must include concepts for expressing object-subobject relationships.
If there are no good means for object-orientation, an alternative may become attractive: to leave the interpretation of the object structure (and, hence, its maintenance and consistency) entirely in the hands of the user. In this case the data model should offer fields of unlimited length into which the user may encode his structure; these "long fields" remain unstructured from the DBMS point of view.

DM2 **Version support.** The data model should offer a flexible version mechanism that allows the user to model the various semantics of versions he needs (some of which we mentioned above).

DM3 **General relationships.** Complex objects and versions can be considered as establishing (special) relationships between objects. Other relationships can readily be identified. Take our VLSI example where numerous relationships have been introduced (Fig. 2.16), not all of them of the object-subobject kind (e.g., the connection of a path to a cell via a port, or the adjoinment of cells). Generalization is another such aspect. Consequently, the data model must contain a mechanism to express general relationships between (perhaps complex) objects.

DM4 **Extended attribute domains.** Attributes reflect the properties associated with an object as a whole. Besides the usual numerical and character string attributes, engineering applications sometimes need support for lists, matrices, and sets as attributes, and for the above-mentioned long fields.

2.3.3 Database System Interface

A database language is a faithful – though not always comfortable to use – reflection of the data model. Database languages, no matter how cleverly designed, cannot add capabilities that are lacking in the data model. But the reverse is also true: A database language should make available all of the features of the data model and not hide any. Unfortunately, the more expressive the data model, the more complicated can the language be expected to be. In particular, the number of DML operators will have to increase beyond what was presented in Sect. 2.2.2.

Fortunately, in design applications the human end-user generally does not operate directly on the database, but accesses the database by using some kind of tool. Consequently, a database interface must be tailored to tools rather than to human users. In general, what is required is a database programming interface which allows database operators to be called from programs (i.e., tools) written in some usually high-level programming language. For quite a long time database systems have been providing both, programming interfaces – usually for languages like COBOL or PL/1 – and interactive query interfaces. In engineering applications modern programming languages like Pascal or Ada will play an important role; nor should FORTRAN be ignored.

The programming interface must reflect the object-oriented nature of the data model, i.e., it must contain all operators necessary for declaring, accessing, and manipulating object-oriented databases. In addition, some kind of interactive query language is useful for quick inspections of the database, but this might be a feature of lesser importance.

UI1 **Object-oriented database programming interface.** Engineering databases must provide for a programming interface which gives the tools easy access to the database in an object-oriented way. In particular, the programming interface must support operations for the manipulation of complex objects and not only for "flat" records or tuples. Since only the latter correspond to "natural" types of programming languages, conversion problems may arise at the interface. Finally, the interface must be structured in a way that allows control of its complexity.

2.3.4 Database Consistency

Even if the user is provided with a more adequate data model than those discussed earlier, explicit consistency constraints will still play an important role in describing the semantics of a miniworld. Examples beyond those of Sect. 2.2 that are typical of design applications are:

- constraints that express how representations of two successive design phases match,
- constraints that express how alternatives and revisions of complex objects may be composed into so-called configurations, and
- constraints expressing technological or physical limitations.

In order to have the database management system enforce consistency, the user must be able to specify consistency constraints. Engineering applications require both descriptive and procedural constraint specification.

- A descriptive form of constraint specification is used for relatively simple constraints. The user states in a predicate or similar form, what the legal database states and/or legal state transitions are, much along the line of Figs. 2.47 and 2.49. The database management system includes test routines that interpret the descriptive specification, or generates suitable test routines for them.
- The procedural form of constraint specification allows the user to build his own test routines himself, and make them available to the database management system (see Fig. 2.48). This method is best suited for constraints that are too complex to specify descriptively, or in cases where the user has already developed some design tools, that include or do checking (like routers and simulators for VLSI design).

It is not sufficient to specify what valid consistency constraints are. It is also necessary to express which constraint is to be checked at what time. Because design processes consist of phases and steps, a particular constraint may not be satisfied until a particular step has been completed. For example, whether or not two paths overlap in a cell design can be checked meaningfully only after the designer has no further intention of moving cells around, i.e., only after he declares the layout completed. One may even claim that global consistency is not reached until the whole design process has been finished. For example, whether the functional design of CELL3 really meets its objectives cannot be stated with certainty until the physical layout has been analyzed in detail. Consequently, the user has to tell the database management system which consistency constraints to check at what time, or, in other words, the database management system must be able to tolerate inconsistent states.

Finally, the user should be in a position to state to the DBMS how to react if a constraint violation has been detected. In traditional database systems the violating transaction is automatically undone by the DBMS. This reaction may not always be adequate, e.g., in those cases where the user does not want the DBMS to destroy his design work, which may have taken him several hours to complete.

To summarize, engineering applications impose the following requirements with respect to consistency:

CS1 **Flexible specification of consistency constraints.** The user should be able to specify consistency constraints both in a descriptive and in a procedural manner.

CS2 **Flexible checking of consistency constraints.** The user should be able to express times or events when consistency is to be checked. For example, checks may be required at the end of database operations or transactions, at periodical points in time, within certain intervals, or on the user's explicit request.

CS3 **Flexible reaction to consistency violations.** The user should be able to state what the reaction to a consistency violation detected by the DBMS should be. It is definitely not acceptable that transactions are automatically undone due to consistency violations.

These requirements, although an absolute necessity for engineering applications, place a heavy burden on research and development. As discussed in Sect. 2.2.3, very few commercially available database management systems provide consistency mechanisms other than primary keys and referential integrity. The reason for this is that consistency checks are expensive operations for the database system to perform at run-time.

2.3.5 Concurrency

Like all databases, engineering databases are installed because of their effect on work integration. Consequently, engineering applications require multi-user-accessible databases. The database management system has to provide for both the necessary synchronization and as high a degree of concurrency as possible.

To determine the units of synchronization in engineering databases is a much more complex task than in classical applications. For one thing, the objects have a complex structure rather than being flat. Secondly, the large number of interrelationships and consistency constraints tend to propagate changes across wider areas of the database. Thirdly, the tolerance of inconsistencies will have an effect. Fourthly, since the majority of design projects are team efforts, the team organization will have to be superimposed on the synchronization. And finally, processing of a particular unit may take quite a long time so that the database management system must deal with long transactions.

The classical pessimistic techniques have been described in Sect. 2.2.4. Is locking not at the same time too strong and too weak a measure to control synchronization in engineering databases? Is it too strong because it treats all cases (team structure, complex objects, and tolerance of inconsistencies) alike? Is it too weak since it still does not capture the propagation of effects across relationships and constraints? Or, to give a further example, in case of long transactions locking would lead to a dramatical decrease of concurrency: The data under operation by one user would be locked for the others for hours, days, or – in the worst case – even weeks. Whether optimistic systems will do much better is still an open question.

CC1 **Concurrency control.** An engineering database management system should provide mechanisms that guarantee correct synchronization under the stronger and weaker requirements mentioned above. In addition, these mechanisms must work well in case of conventional (simple object, short) transactions.

2.3.6 Database Recovery

In classical database technology, the concept of recovery is closely associated with that of transaction: The effects of all transactions already completed on system or media failure are to be reproduced, those of the still active transactions are to be undone. The redo part of this concept is a requirement for technical applications as well, whereas the undo part may cause some discomfort in case of long transactions.

Certainly, the user will not accept that an almost finished long transaction is aborted due to some failure – with the effect that some hours or days of work are lost. What one needs is the possibility of having the database management system save on demand the data under operation. The demand may be determined automatically, e.g., by fixed time interval, fixed volume of database changes, fixed time of day, or it may explicitly originate from the designer. Undo may then turn to the state of the last savepoint. This includes the capability of a database management system to undo long transactions completely and thus to reach the state at the beginning of a – possibly erroneous – design step. We summarize:

RE1 **Savepoint recovery.** In addition to the classical transaction-based recovery, the user should have the capability of explicitly or implicitly choosing savepoints. The DBMS must restore a transaction to any one or at least the most recent of its savepoints.

2.3.7 Transaction Management

Classically, transactions are considered units of consistency, recovery, and synchronization. The preceding requirements analysis has shown that engineering transactions cannot be characterized in the same way, although all three aspects still must somehow be dealt with. The consequences are dramatic for database technology because of the central role played by transaction management in database management systems. The notion of transaction – and, hence, the mechanisms in transaction management – must be drastically revised with respect to the requirements on consistency, concurrency, and recovery.

2.3.8 Miscellaneous Services

In many traditional applications protection plays an important role. This role is often underestimated in engineering applications, although the reasons are equally valid:

- Information today is considered as a resource equally as valuable as finances to an enterprise. Consequently, an enterprise must be able to protect its database against unauthorized access, tampering, or destruction.
- Protection mechanisms may to some degree prevent legitimate users from unintentionally doing erroneous operations on a database.
- In design applications, protection mechanisms may be used to guarantee controlled cooperation of the members of a design team. For example, an object that is currently being designed is protected from any access other than by the designer ("private" object). If the design step is finished the object is made "semi-public", i.e., accessible to other team members. At

some later time the object may finally become the "public" property of the enterprise, accessible by, e.g., designers in other teams, and maintenance personnel.

All of this does not sound much different from traditional applications. Consequently, it is only a matter of proper utilization of known concepts to meet the requirement:

M1 **Design-oriented protection concept.** The database management system must provide a protection concept that allows the user to protect objects in ways that mirror their structure, and the entire design organization and process.

In the future, the number of designs will continually grow. Typically, the design data will be needed for many years because of maintenance, error correction, functional extension, optimization, or adaptation. On the other hand, the likelihood of access will decrease over the years so that there is, after some time, no need to keep the data online. Therefore, archiving is of particular importance to engineering databases. In this case the database management system must keep information on which objects have been archived, and where (archive catalogue), in order to reload them on request.

M2 **Archiving facility.** An engineering database management system must provide facilities for archiving and reloading design objects, where the catalogue of what has been archived should remain accessible online.

Again, existing techniques should be able to cope with this requirement, although new storage technologies such as optical disks will have a massive impact.

Engineering applications include intellectual activities to a higher degree than classical business applications. From the database management system point of view, this is reflected in a much less stable environment. Hence, the database schema cannot be considered time-invariant. Facilities must be provided to modify existing database schemas – not least by adding new types. On the other hand, in cases such as integration of a new tool or changes to an existing tool, the tools that remain unaffected should not be forced to take notice of it, which is best supported by providing a rich variety of subschemas or views.

Moreover, the schemas of engineering applications are more complex than those of business applications. (The latter are characterized by a small number of types with a large population per type, whereas engineering applications typically behave the other way round.) Consequently, a need for frequent inspection of schemas is to be expected. Hence, the schema information should be stored in a way that allows the user to formulate queries on it. Separate facilities for the flexible management of schema information, i.e., a data dictionary, are particularly important. This data dictionary should contain, in addition to the schema information, defined consistency constraints, access rights, a history of schema modifications, subschemas or views, usage statistics, and so on. On the other hand, the classical distinction between database and dictionary becomes somewhat blurred because information maintained in classical applications within dictionaries (e.g., version histories) is part of the database in engineering applications.

M3 **Integrated data dictionary.** Flexible data dictionary facilities that are easy to handle by the engineering user should be provided together with or as part of the database system.

2.3.9 System Organization

Design applications usually need rather diverse hardware resources such as:

- high-resolution graphical workstations for interactive design;
- mainframes with high computational power for solving numerical problems (e.g., placement algorithms, routing algorithms, finite-element computations, simulations, and so on); and
- large mass storage for integrating design databases and for archiving them, possibly served by separate computers.

Consequently, database functions must be available on a number of computers that can be assumed to be connected by some network. The whole database is distributed among several computers. Distributed databases (Fig. 2.63) appear to be a desirable long-range goal. Present technology for them, however, still has difficulty coping economically with consistency, concur-

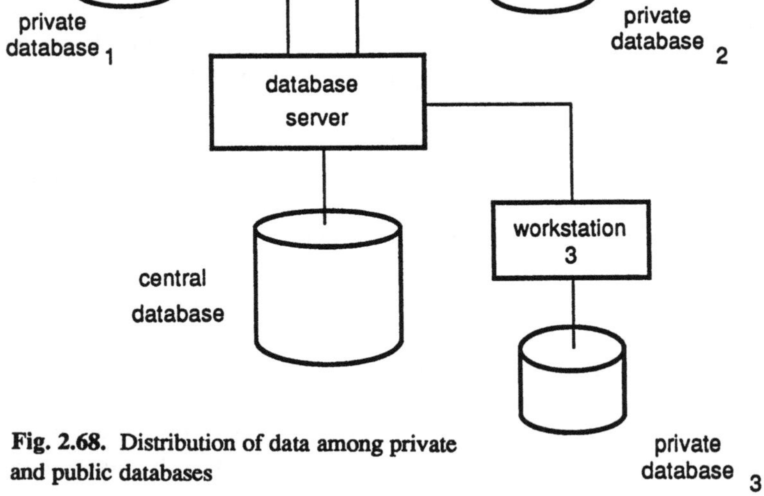

Fig. 2.68. Distribution of data among private and public databases

rency, and recovery issues. Hence, a meaningful alternative is to make one computer a central database server to which all others are attached as clients (Fig. 2.64).

Users are expected to do their design work at the workstations. Before work begins, the object of interest will have to be transferred from the central database to the (private) workstation database, and back after completion of work (Fig. 2.68). The database management system should provide operators ("checkout" and "checkin") necessary for this transfer. To sum up :

SO1 **Multi-database support.** The database management system must support the processing of multiple (private and public) databases.

SO2 **Server-client architecture.** The database mangement system must support databases that are distributed among several computers, where these are organized in the form of one server and multiple clients. Operators for transfer of objects between different computers are required.

SO3 **Remote database calls.** A facility to execute remote database calls – i.e., to call database operators on a remote computer – is required.

2.3.10 Performance Control

In the past, performance (or lack of it) has been a crucial point in applying database management systems to engineering applications. The requirements for data models have shown that one should be able to specify to the database system which objects are to be considered a unit for a particular application (complex objects). In turn, the database management system should make use of this information such that objects are processed particularly efficiently if they correspond to specified units, e.g., by storing the subobjects in adjacent areas of peripheral storage (clustering). In addition, access paths must be generated that support the Braphical-interactive way of object manipulation.

Another mechanism that promises performance enhancement is an object-oriented main memory database. The database management system should allow the user to express what complex object he is going to work on, in order to load the whole object into main memory and to avoid the disk accesses otherwise needed during the period in which the object is processed. Interactive applications with high read-access data rates may gain considerably in performance from this technique.

PC1 **Enhanced clustering and access path strategies.** In order to store complex objects in physically adjacent disk areas, enhanced clustering strategies are needed. Rapid associative access is to be supported by a variety of access path strategies, some of which must be tailored to interactive and graphical object manipulation.

PC2 **Object-oriented main memory database.** On specific request a complex object should be kept in main memory for the entire period of operation. Manipulation of the complex object, however, should not differ functionally from that of objects accessed on peripheral storage.

Access paths supported by present database technology usually are based on hash or B-tree techniques for assocative access, and on pointers for structural access. It remains unclear to which extent they meet the needs of the manipulation of complex objects. A vast stack of more complicated access techniques is found in the literature; some of these should be evaluated further as to their suitability to the purpose at hand.

The clustering capabilities of present database systems are generally very limited. They only cope with cases where no clustering conflicts occur, as for instance clustering of the set of tuples of a single relation, or clustering of a set in the network model. Clustering of complex objects that share some subobjects is not supported at all.

Finally, buffering techniques are usually based on the unit of peripheral storage access: pages or blocks. The object structure is not taken into account.

2.4 Data Organization and Manipulation: Taking Care of the New Focus

2.4.1 Trends and Objectives

As we have seen from Sect. 2.2, database management systems that are currently available in the marketplace are by no means useless for engineering applications. On the other hand, they are less than optimal for that purpose, since none of them meet all the requirements listed in Sect. 2.3. One objective of modern database research is to overcome some of the shortcomings and meet more and more of the new requirements. Indeed, during the past couple of years an increasing number of researchers have attempted to develop concepts and prototype systems that are better suited to these requirements.

In this section, we present some of the new approaches. We organize the presentation again along the lines of Sect. 2.2 and 2.3.

As will be seen, the data model issue shows the largest variety of proposals. This is certainly due to the fact that the information modeling and manipulation facilities form the major part of the DBMS interface and thus represent one of the most pressing deficiencies of traditional solutions. It is, however, also true that some of the other database concepts for engineering applications are currently not as advanced as data models are, or they simply do not lend themselves to a large variety of new approaches.

Note that in practice all of the features of the systems presented here are prototypes that have been built to demonstrate one concept or another. While some of them will certainly remain research vehicles, there is little doubt that others will get the necessary improvements to let them compete on the market in the not too distant future. It is, therefore, not hard to predict that more and more of the concepts we discuss below will show up in commercial products, in one form or another.

The intent of this section, then, is to give an overview of current developments in order to enable the reader to judge for himself what he might expect from systems of the future. He should, however, be warned that details will still change (in fact, we have taken the liberty to modify some of them slightly for better presentation within this text).

2.4.2 Data Models

2.4.2.1 Semantic Concepts

Semantic concepts provide high-level mechanisms to express the information of an application area within a structural framework. They are usually tailored to a particular class of applications and thus too rigid to be used directly as a DBMS data model ("semantic" concept or data model is a well-established notion although not very precise: Any model expresses *some* semantics, and "semantic" models are just thought to express *more* than others).

The purpose of introducing a set of semantic concepts here is twofold. For one thing, we need to have some framework available for model comparison that is independent of the intricacies of specific data models. For another, there has been a tendency in the past to fashion the "classical" semantic concepts into a data model of their own because they overcome some of the deficiencies of the classical data models – although the concepts usually cover structural aspects only, and do not include domain types or a comprehensive set of operators.

Entity-Relationship Approach to Data Modeling (ER)

Since its first publication some ten years ago, the ER model (ERM) has been widely used as a semantic model. Though not specifically conceived for engineering applications, it forms a good starting point for any kind of data modeling. Moreover, the pure ERM has meanwhile been augmented by a wealth of additional concepts to express more semantics. We will discuss only a few of them.

The basic assumption of the ERM is that the miniworld consists of a number of distinguishable *entities* (which may be concrete things like a house or a person, or abstract things like an idea or a project) and *relationships* between those entities. Entities are characterized by the values of their *attributes*. Entities are either *weak* (if they exist only in a relationship with some other entity) or *strong* (if they exist independently of others). We remember that the semantics of relationships need more expressive power than that provided by either the network or the relational data model. In the ERM relationships are n-ary, and entities play certain *roles* within a relationship. Like entities, relationships may have additional descriptive attributes (beyond role attributes).

Both entities and relationships are instances of *entity types* or *relationship types*, respectively. All instances of a given type are equipped with the same attributes. With each type a class is associated. A type describes the set of all

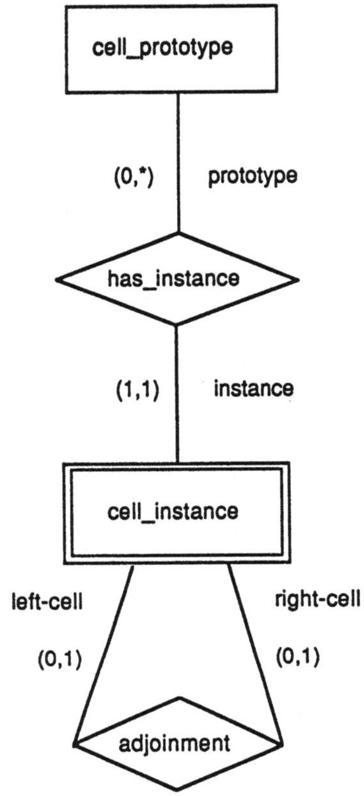

Fig. 2.69. Example of an entity-relationship schema (with attributes omitted): boxes represent entity types (double lines: weak entities); rhomboids represent relationship types; "left cell", "right cell", etc. denote roles in relationship types; minimal and maximal cardinalities are given for each role (e.g., "every instance has exactly one prototype", "an arbitrary number of instances may exist for every prototype", "every instance has at most one left or right neighboring instance")

possible entities/relationships with common properties, while a *class* comprises all instances that currently exist in the database.

ER diagrams are a convenient means of depicting the interpretation of a database state or – more frequently – of a database schema (in the latter case the diagram is in terms of entity or relationship types). Figure 2.69 gives an example.

Still, expressive power has not been increased by much. For example, relationships are general in nature. If a relationship is to be more restrictive, e.g., purely functional or even bijective, we need additional means of expression. Adding *cardinalities* to relationship types is such a means. A cardinality expression is a pair (mincard, maxcard) indicating how often an entity must at least and may at most participate in a given role of a relationship of the respective type. Figure 2.69 gives an illustration. Note that the notion of strong and weak entities is essentially a way of expressing the dynamics of relationship types; weak entities are somewhat akin to the FIXED AUTOMATIC option of the network model.

We remember from Sect. 2.3.2 that the abstraction mechanism of generalization is another of those concepts of importance in modeling engineering applications. The ERM includes such a concept as well as a concept of aggregation (for examples, see Fig. 2.70):

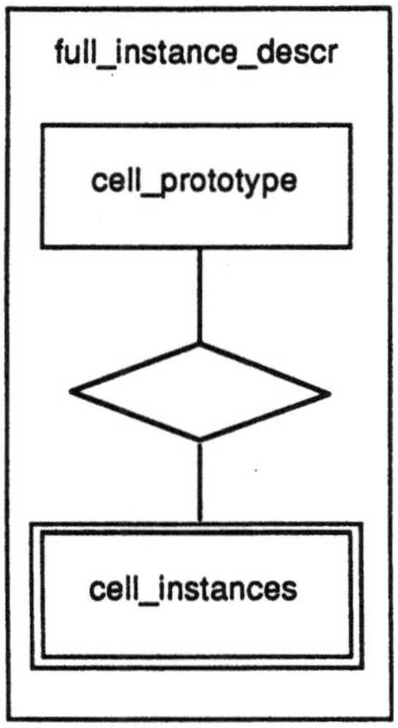

Fig. 2.70a

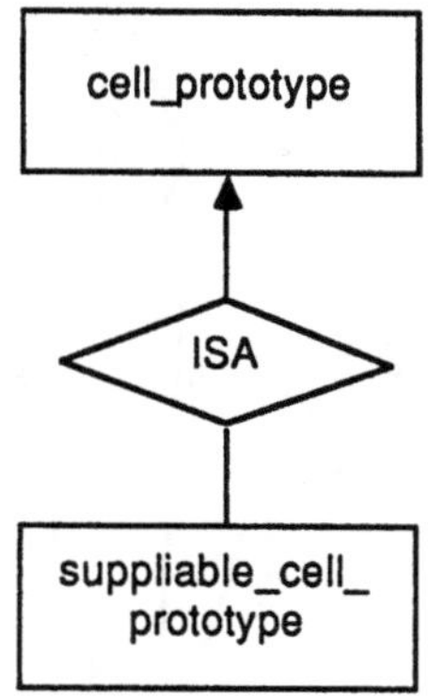

Fig. 2.70b

Fig. 2.70a. (Atomic) Aggregation: an instance of a cell together with its related prototype (and the relationship between them as well!) from a new entity (which may again enter into relationships, etc.)

Fig. 2.70b. Generalization: the special relationship ISA creates a subtype/supertype structure. Subtypes inherit the attributes of their supertypes. For example, the description of a suppliable cell prototype comprises all attributes of a regular cell prototype, but will have some additional ones (for example the supplier name and the order number)

- *(atomic) aggregation* regards a relationship among entities as a higher-level entity in its own right, where the details of the lower level are suppressed; and
- *generalization* views entities that are instances of one type to be, at the same time, instances of a second type (subtype and supertype). Usually this makes sense when a number of subtypes generalize to one supertype containing the attributes that all subtype instances have in common. For example, types "cylinder" and "cube" may be generalized to supertype "volume" with attributes "material", "color", and so on. "Cylinder" instances have additional "height" and "Biameter" attributes, while "cube" instances are characterized by a specific "length" attribute. Generalization leads to the concept of *inheritance*: Subtypes inherit the properties (i.e., the attributes in the simplest case; there may also be operation inheritance) of their supertypes. *Multiple inheritance* occurs when a subtype has more than one (direct) supertype, and of course generalization may extend across more than one level of a hierarchy. The concept of generalization is often referred to as the IS-A relationship ("every cylinder *is a* volume"). The opposite view is called *specialization.*

With these concepts in mind we attempt to model our VLSI example. Figure 2.71 shows one conceivable result that makes use of ERM concepts (it is still simplified in that it does not take the external ports of the enveloping

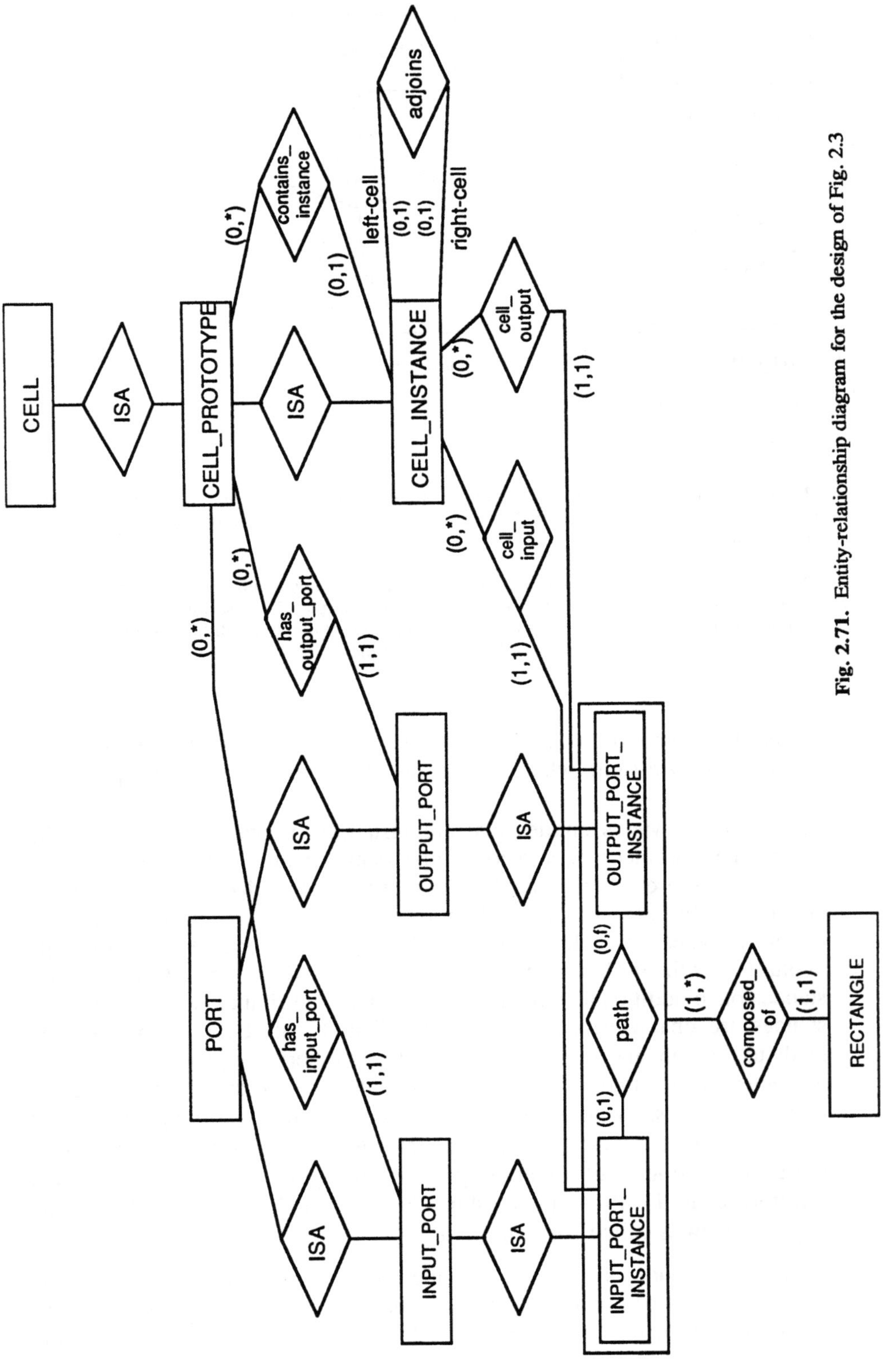

Fig. 2.71. Entity-relationship diagram for the design of Fig. 2.3

cells into account). The reader may notice that the design reflects all of the topological constraints (rules 1, 2, and 3) but nothing else. This supports our previous contention that even with richer data models consistency constraints cannot be done without.

Figure 2.71 demonstrates rather well the role of semantic data models in the design of database schemas. Requirements analysis should be expressed with the aid of such a model. There are mapping rules that translate ER schemas into network schemas or relational schemas, preserving as much of the semantics as possible. These rules also clearly identify which semantics are lost during the mapping and must be recovered in terms of consistency constraints, or as part of application programs. As an exercise, try to construct a network schema from Fig. 2.71.

Molecular Objects

The representation of complexly structured objects has been found to be one of the key issues in engineering database systems. Though it might be used in general, the concept of atomic aggregation as introduced above is not very convenient to express object composition. Therefore, the additional concept

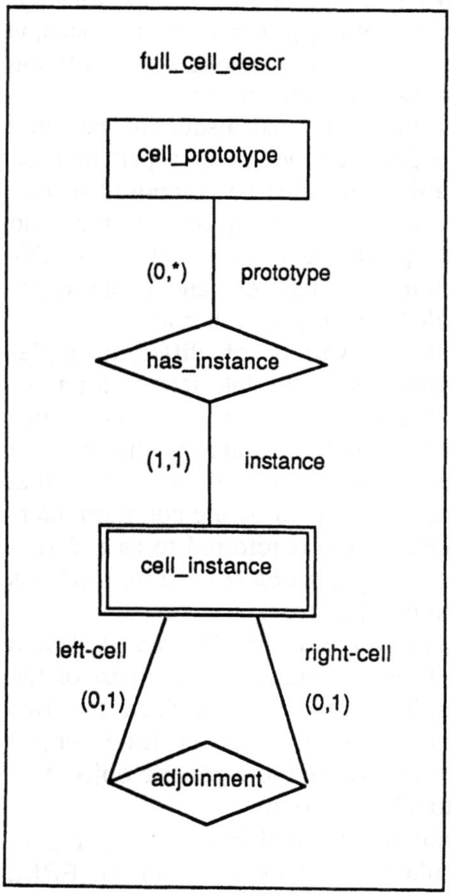

Fig. 2.72. Molecular objects: A number of cell instances and prototypes together with their relationships are combined into a new molecular object of type "full_cell_descr". Thus, the layout structure of Fig. 2.3 can be described as a molecular object of this type (by adding path information, etc.)

of *molecular aggregation* has been introduced: It aggregates an arbitrary set of entities and their relationships into a higher level entity (a *molecule* or *molecular object*) and thus allows the expression of arbitrary compositions of entities including selected relationships among them. Of course, molecular objects may themselves participate in arbitrarily defined relationships in the same way as atomic entities do. Figure 2.72 illustrates the concept.

Molecular objects may be *disjoint* (if they do not share any subobjects/relationships) or *non-disjoint* (i.e., they overlap in some part). Also, they may be *recursive* (i.e., a molecule is composed of other molecules of the same type) or *non-recursive*. There are thus four classes of molecular objects, ranging from the very general non-disjoint/recursive to the much simpler disjoint/non-recursive ones. We may use this classification to evaluate the complexity of objects and the power of systems that manage them.

Design Objects and Versions

A number of other semantic concepts have been proposed that allow specification of even more semantics of an engineering application. Obviously, the more tailored a concept is to one specific application area, the less useful it is for others. Though specific concepts may be of great use for their particular field, we content ourselves with presenting just three more concepts of broader applicability at least in the area of design management: version generalization, version instantiation, and version parameterization.

To approach the subject, we start with the traditional distinction between the interface and the imBlementation of a designed system or a part thereof. A molecular object representing such a design will thus be composed of two subordinate objects at the uppermost level, one representing the interface and one the implementation. Consider the cell prototype in Fig. 2.71 where INPUT_PORT and OUTPUT_PORT represent the interface, and contains_instance in conjunction with CELL_INSTANCE the implementation.

For a given interface there may exist a variety of different implementations; each such implementation together with the interface is a particular entity of type CELL_PROTOTYPE, that may in turn result in any number of instances of type CELL_INSTANCE. What is apparently missing is a concept that expresses the fact that there are subsets of cell prototypes that coincide in their interfaces, with each subset represented by the common interface. In some of the literature such a representation is referred to as a *design object type*, and a member of the subset as a *design object version*. Versions *inherit* the interface of their respective design object type.

In fact, in design applications a more general view should hold: A design object type contains all aspects common to the versions, be they part of the interface or the implementation (indeed, the distinction between the two should be entirely dropped). The term *generic object* instead of design object type is thus more appropriate. Versions are generalized to a generic object that is an abstraction of the features common to all its versions.

Notice that the concept of version is not yet included in Fig. 2.71. Figure 2.73 indicates what the version idea would have to look like in the ERM.

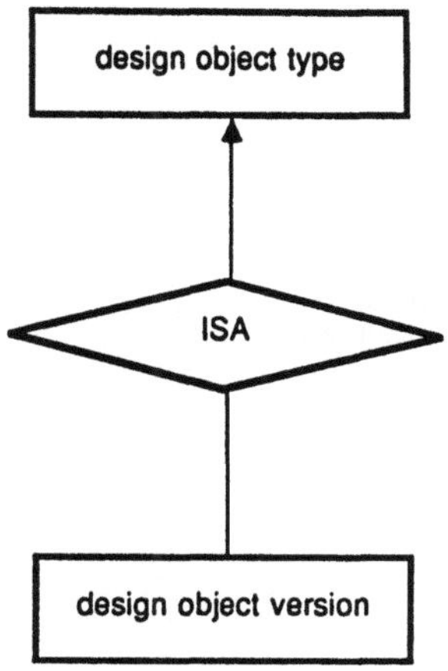

Fig. 2.73. Version generalization expressed by ER concepts

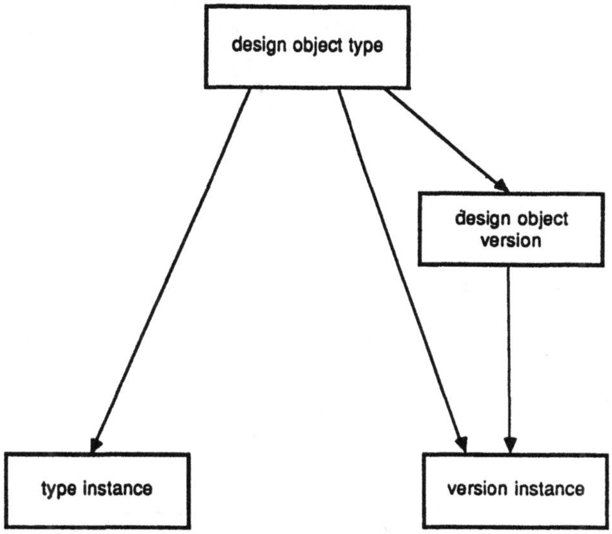

Fig. 2.74. Inheritance graph for types, versions and instances (→ : inherits attributes)

Notice also that Fig. 2.73 cannot be integrated at all into Fig. 2.71, because the generic object is not a type in the sense of Fig. 2.71 but an individual item. Again we conclude that what we need – moving across different levels of abstraction – cannot be provided by the ERM and would necessitate further semantic constructs.

In fact there are even more semantics to versions. Once a generic object has been defined, corresponding design objects may be used by incorporating them into more complicated designs. This necessitates a more discriminating concept of *instantiation*: A generic object is instantiated by copying its interface only, while in the instantiation of a design object version, both the interface and the implementation are copied (Fig. 2.74). In either case, the instantiation inherits the attributes of its generic object/version. However, there may be further attributes that are specific to the particular instantiation (e.g., the placement coordinates of a subdesign in a comprehensive one).

The use of generic objects, versions, and instantiations may further be refined to the concept of *parameterized versions*. In an implementation of a design object, we usually require that a specific version instantiation for subobjects has a complete specification of the current design. However, it is often convenient or even necessary to postpone the decision of what version to take, and to include just an instantiation of the generic subobject (i.e., the interface). We thus obtain an object version that is incomplete in the sense that the "sockets" defined by generic object instantiations remain to be "plugged" by choosing a concrete version thereof later on. Parameterized versions thus act as *templates* from which various *configurations* may be constructed. For an illustration, see Fig. 2.75.

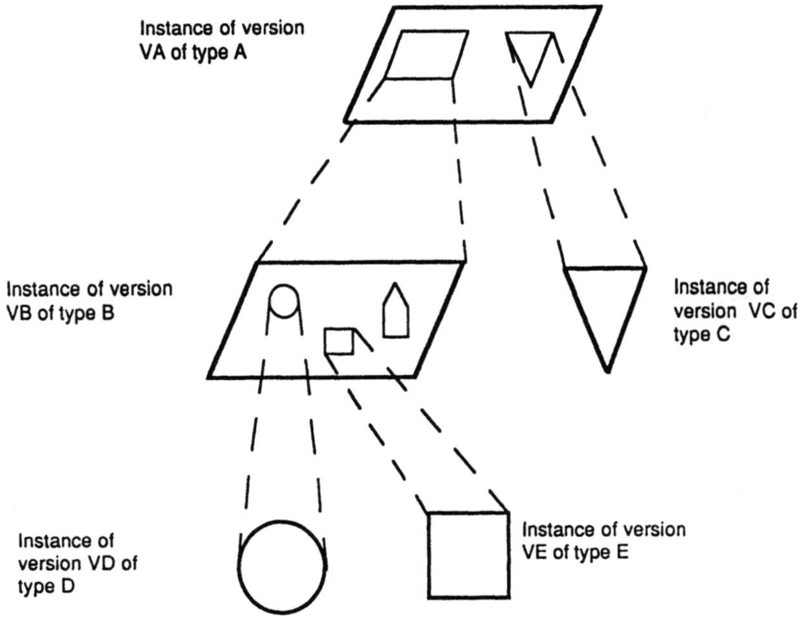

Fig. 2.75. Parameterized version: the instance of version VB of type B is parameterized by one parameter, as no version has been plugged in so far for the subobject of type

2.4.2.2 Object Identification

Any proposal for a data model that is supposed to capture complex engineering structures has to deal with the problem of how to identify the entities of interest. In conventional database applications, this is (at least as far as the system interface is concerned) completely left to user-defined keys. At best, the database system software guarantees that these keys are unique within some part of the database at any point in time.

A number of weaknesses can be traced to user-defined keys, some of them even in traditional applications.

- Keys are rather artificially chosen (consider, e.g., an employee number that is by no means a natural property), or they tend to change during the lifetime of the entitiy they are supposed to identify. In this case – besides other problems – the fact that an entity is still the same is completely lost.
- When two databases have to be merged or when they are used cooperatively, it is most likely that conflicts between identifiers will arise.
- When a database comprises a large number of entities, it becomes very tedious and uncomfortable for the user or his application programs to be responsible for the generation and assignment of unique identifiers.
- Finally, consider that database objects are manipulated at a terminal screen via graphics software – a situation typical for engineering applications. The way of referring to an object in such an environment is certainly not by stating any sort of identifier, but by pointing a mouse or some similar

device to a specific screen position. Thus the user may consider identifiers a purely technical issue.

All these problems suggest using *surrogates* for object identification. A surrogate is a value that uniquely identifies an object and is system-generated, never changes during the lifetime of the object, and is not reused after deleting the object.

Users may thus exploit object identifiers (and rely on the above-mentioned properties), but of course cannot modify them. Note that user-defined keys may still exist, though they are not necessary. Functions that translate a surrogate into an associated user-defined key or vice versa may then be provided.

The generation of surrogates is usually done by concatenating as many of the following parts as are necessary to attain uniqueness within the area of interest (ranging from "worldwide" to "local database"):

- processor identification,
- time of object creation,
- object type,
- sequence number (per type, per database), and
- database identification.

Despite their well-defined composition, surrogates are usually presented to the user without associating any semantics to them. In order to store surrogates with the objects they identify (or for whatever other purpose), the set of attribute domains has to be extended by a special type *surrogate*.

2.4.2.3 Objects and Relationships

The mechanisms that cast the application semantics into general objects and relationships between them add much to the usefulness of a database system for engineering applications. It is, therefore, little wonder that numerous proposals have been made for including them in data models. They usually comprise:

- molecular objects (termed complex objects, structured objects, or the like),
- relationships between objects (with varying flexibility), and
- object versions (commonly conceived less strictly than in Sect. 2.4.2.1).

In order to be suitable for a broader range of applications, database systems tend to avoid the incorporation of concepts for very specialized semantics. Consequently, the latter (including some of the cases in Sect. 2.4.2.1) must still be expressed in terms of more general concepts.

We shall discuss below what form these proposals take in some of the modern approaches to engineering data models. With respect to their conceptual origin, we distinguish the following categories:

1. extensions to conventional relational systems,
2. unnormalized relations, and
3. systems with explicit structuring capabilities.

We shall give an example for category.

Extended Relations: XSQL

XSQL extends the classical SQL/DS (formerly System R) relational concepts of IBM by – among other things – hierarchically organized sets of tuples from multiple relations, called *complex objects*. Structurally, this is achieved by adding two new attribute types, IDENTIFIER and COMPONENT_OF. IDENTIFIER columns hold (worldwide unique) surrogates, while COMPONENT_OF columns are used to organize a tuple hierarchy in the following way:

- A complex object consists of one root tuple (taken from a root relation) and an arbitrary number of tuples of one or more other relations (called child relations).
- Every relation participating in complex objects has exactly one IDENTIFIER column. The identifier in the root relation serves as the surrogate for the complex object.
- Every non-root relation has exactly one COMPONENT_OF column. A tuple value in it identifies the parent relation tuple of this child tuple. Note that one child tuple has exactly one parent tuple, but multiple child tuples may have the same parent tuple. The restriction of one COMPONENT_OF column per child relation guarantees a strict tuple hierarchy.
- Complex objects have to be inserted tuple by tuple in a top-down fashion: Insertion of a root tuple creates a new complex object, rendering its IDENTIFIER value which can afterwards be used as the COMPONENT_OF attribute value of child tuples, and so on (provided, of course, that the parent tuple relation has been specified for the COMPONENT_OF column).

So-called *references* are used as an additional concept to represent non-hierarchical structures as well as relationships between complex objects. INTERNAL_REFERENCE attributes are used to refer to one tuple from another within the *same* complex object, and EXTERNAL_REFERENCE attributes allow reference to one complex object from another. Both references are one-way only, and are established by exploiting the identifier of the tuple to be referenced.

As for operations, XSQL provides facilities to copy complex objects and to delete entire complex objects or subobjects (subtrees of the hierarchy). For example, the deletion of a root tuple automatically deletes all of its (direct and indirect) child tuples. On the other hand, references are currently not subject to any implicit consistency control, so that there is no check whether they point to an object/tuple that has been deleted. The FETCH operation allows retrieval of any meaningful subset of the tuples of a complex entity (e.g., restricted to specified attributes or attribute values, a limited number of levels in the hierarchy, tuples from specific component relations, etc.) and stores the subset in host language structures. While there is no interactive query language provided for complex objects, all pure relational concepts may be used (thus a complex object can still be retrieved piece by piece by applying interactive SQL).

Figure 2.76 tries to model our standard example by means of XSQL complex objects. It demonstrates how the hierarchical decomposition of a cell object into instances of other (predefined) cells and paths as well as of a path into rectangles is reflected by making INSTANCES and PATHS tuples children of a tuple in the root relation (CELL), and so on. The real world fact that instances relate to a prototype cell (which is different from the one they are themselves a part of) is expressed by an external reference from an

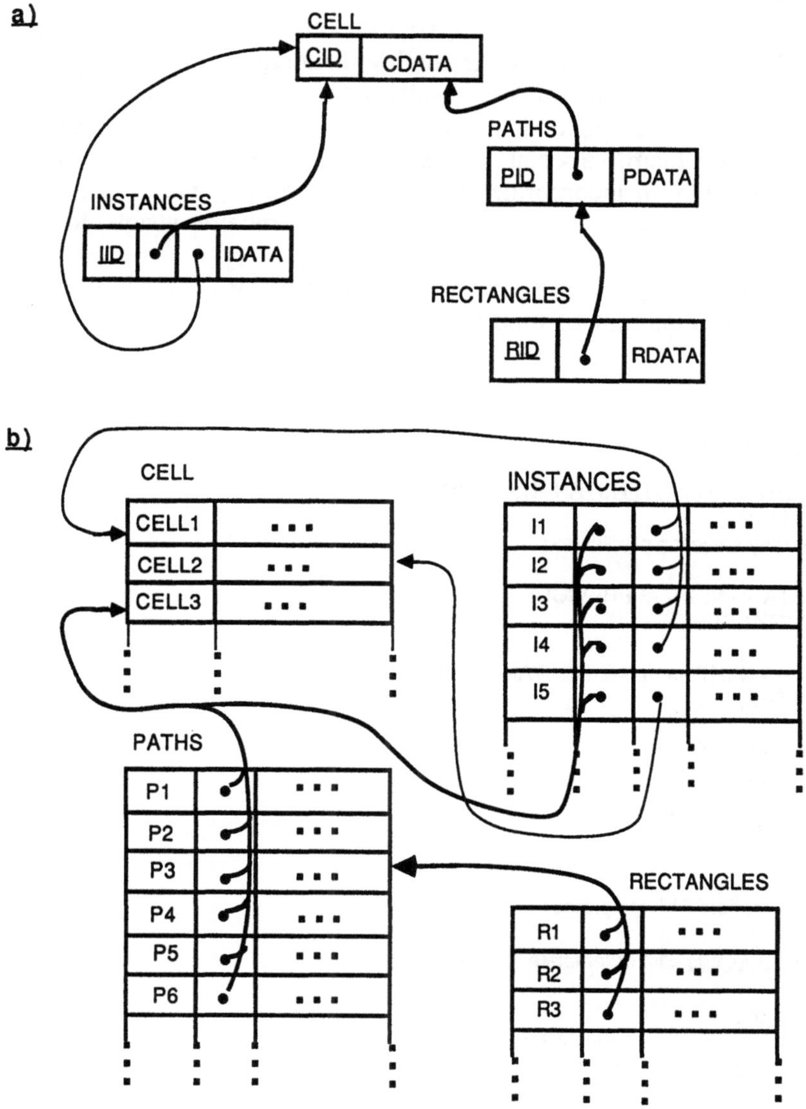

Fig. 2.76a,b. Graphical representation of an XSQL database for a composition according to Fig. 2.7. Heavy arrows indicate COMPONENT_OF attributes, light arrows refer to the ISA-relationship of Fig. 2.71 and indicate (external) REFERENCEs. Underlined attributes are of type IDENTIFIER. Other design data have been omitted. **(a)** Schema. **(b)** Example database

INSTANCES tuple (of one complex object) to a CELL tuple (of a different complex object).

As COMPONENT_OF attributes may never have NULL values, instances, paths, and rectangles can only exist if the appropriate cell exists. Since COMPONENT_OF values point upwards in the hierarchy, the reverse is not true: The system does not guarantee that instances whose prototype cells have been deleted are also removed from the database. Thus, incomplete data may result (instances may need the respective prototype cell information be used in any meaningful way).

Figure 2.77 shows a sample FETCH operation on the above database using XSQL embedded into the PL/1 programming language. Note the typical sequentialization of (potentially) sets of results by means of a cursor concept. If a *where* clause was specified, it could be used to restrict the qualifying objects, or parts of them. Special implicitly defined attributes (->FIRST, ->NEXT, etc.) have been incorporated into this specific application program interface of XSQL to make the hierarchical tuple structure visible to the programmer.

The design of XSQL aims at providing useful extensions for engineering applications with a minimum of modifications to the existing SQL relational

```
exec sql declare cc complex cursor for

   select CID, CDATA, -> FIRST(INSTANCES), -> FIRST(PATHS)
          from CELL
          where <any selection predicate>;

   select IID, IDATA, -> NEXT, -> PREVIOUS
          from INSTANCES;

   select PDATA, -> NEXT, -> FIRST(RECTANGLES)
          from PATHS;

   select RDATA, -> NEXT
          from RECTANGLES;

exec sql end complex cursor;


... definition of appropriate PL/I templates ...

exec sql open cc templates <list of template identifiers>

if sqlcode = 0 then do;
   exec sql fetch cc into <template identifier>;
   if sqlcode <> 0 then call <error handling>
end;

exec sql close cc;
```

Fig. 2.77. Sample PL/1 application program with embedded XSQL to retrieve data of cells according to the schema defined in Fig. 2.76a

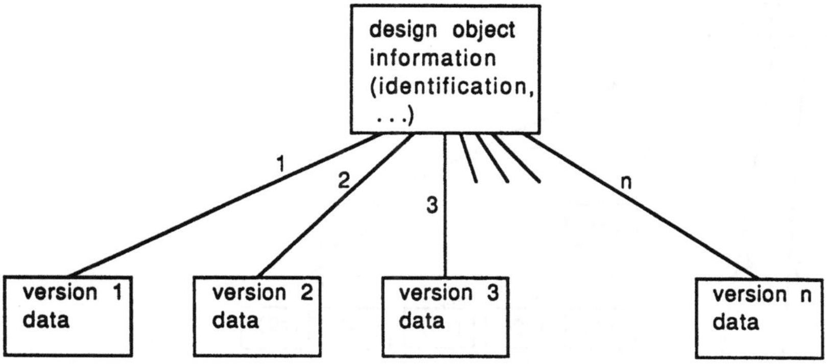

Fig. 2.78. XSQL design object with versions

system. While this has been achieved with regard to the data model itself (only a few new column types and operations had to be added), major efforts have been necessary to change the system optimizer which must know about complex object structures to be able to support efficient processing of operations on them.

XSQL also provides mechanisms to support a basic version concept. Versioned objects are called *design objects* and consist of a generic part (to represent information common to all versions) and one complex object (of the same type as all other complex objects) per version (Fig. 2.78). Versions are created individually (either from scratch or by copying a previous version) and are numbered sequentially within their respective design object.

Regarding the object-oriented way of working in design environments, the interface of the version concept in XSQL is a purely operational one, i.e., versions cannot be defined as such in the schema. Consequently, during database design, the user defines the complex object type he wants to use for the versions of a design object like he defines any other complex object type. At runtime, a design object may be created by choosing an existing complex object type for its versions. Within a design object, versions may be selected for retrieval, update, or deletion by their number. Versions may be frozen and thus preserved from further modifications. The "current" version (i.e., the one with the currently highest version number) and the "last frozen" version (i.e., the frozen one with the currently highest number) of a design object may be selected without knowledge of their number. Of course, reference to a version may also be made via the surrogate of the complex object representing it.

The current implementation of versions is completely on top of the XSQL interface. Consequently, version operations have not been integrated into the query language. In most cases, this will not cause any inconvenience to users: They would first select the desired version and afterwards use conventional XSQL to query or to manipulate it in the same way as any other complex object.

For the purpose of arranging a potentially large number of versions within a design object, the concept of *version clusters* has been introduced. If we

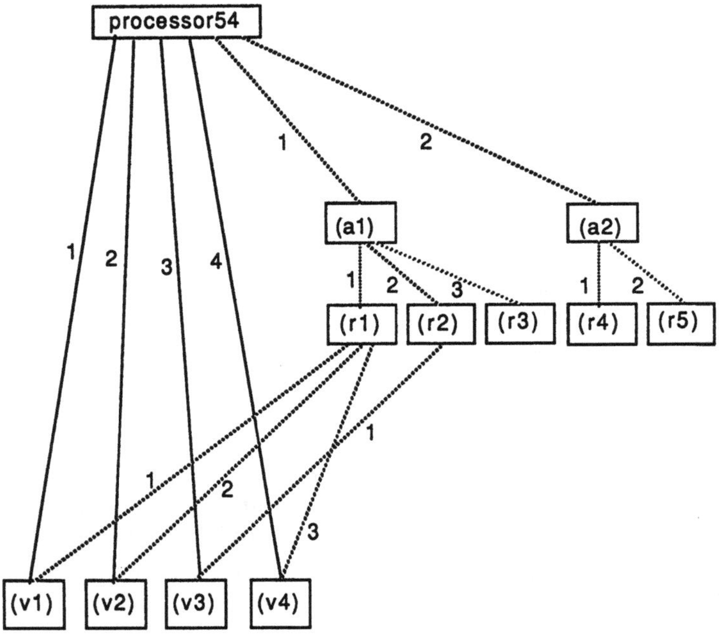

Fig. 2.79. A design object with 4 versions and intermediate version clusters: e.g., (v4) may be reached via top level cluster 1 (= a1), next level cluster 1 (= r1) and from there via version number 3. Alternatively, it is version number 4 when not considering the cluster structure. (possible interpretations: a: alternatives, r: revisions)

view a design object as a collection of all its versions, then version clusters are a means of grouping arbitrary subsets of versions or lower level clusters into units that may be referred to below the design object level. Figure 2.79 shows an example design object where a number of versions have been grouped into "revisions" and a number of revisions form an "alternative". The design object itself, on the top-most level, consists of a number of alternatives. The cluster organization is hierarchical, but not necessarily tree-like.

Cluster structures are superimposed on the standard arrangement where versions are reached via their sequence numbers within the design object. Now, every version or subcluster within a given cluster is assigned a sequence number relative to its ancestor which may be combined to form logical access paths that form a thread through the cluster structure to the desired cluster in a version.

Clusters can be created and deleted dynamically, and the membership of versions and clusters in (super-)clusters may be changed at any time. However, the types of admissible clusters and connections between them have to be defined together with every design object type. In our example, we have clusters of type "revision" and "alternative" with connections as explained above. Thus, it would be impossible to make a version a member of an "alternative" cluster or to create a "revision" cluster without indicating the "alternative" cluster in which to include it. Like design objects, clusters may bear information that pertains to all their members.

Finally, the version mechanism of XSQL addresses the problem of interobject references (i.e., external references) in the presence of versions. Consider a reference from object A to object B. If B is a design object with versions, it has either to be decided during the construction of A which versions of B should be taken, or versions of B should be dynamically reassigned to A. This is impractical for the many situations when references should be automatically rerouted in case new target versions become available. XSQL thus offers the concept of *generic references* that do not decide which versions to take but indicate the desired object type only. The resolution of a generic reference to a specific version is postponed until runtime and is based on so-called *environments*. They contain information on which version to take for which generic reference that may be provided on a user- and/or source object-specific basis, and that can be defined very conveniently by resorting to a number of frequently required defaults. [Dittrich and Lorie 88] contains more details and information.

Unnormalized Relations: NF^2

A second way of introducing molecular objects into the relational world is to drop the standard assumption of first normal form (1Nf). First normal form requires that all attributes of a relation be atomic from the DBMS point of view. If one allows non-first-normal-form (NFNF= NF^2) or nested relations, an attribute can itself be relation-valued, so that hierarchically organized molecular objects can be expressed. Consider Fig. 2.80 which models our running VLSI example. While CID and CDATA in relation CELL are supposed to be conventional "flat" attributes, PATHS and INSTANCES are attributes that are themselves relation-valued. Thus, one CELL-tuple contains an entire (sub-)relation as the value of its INSTANCES attribute. As can be seen from the PATHS attribute, relations may be nested across multiple levels.

NF^2 is the data model incorporated into several experimental systems. Like XSQL, its disadvantage is that non-hierarchical aspects have to be expressed in a different way than hierarchical ones. The merit of both approaches is that they stay more or less within the relational world. This is particularily true for NF^2. Relation-valued attributes may be treated in the same way as relations. Thus, only slight modifications of existing relational query languages (e.g., to account for nested queries) are necessary. Moreover, NF^2 relations are easily transformed into flat ones, and vice versa.

One of the prototype systems implementing the NF^2 data model is the AIM prototype of the IBM Scientific Center in Heidelberg. Beyond relations, it also supports lists and multisets as attributes. Query operations include:

- tuple-specific operations as usual,
- list-specific operations (concatenation, sublist extraction, and accessing elements by subscripts),
- multiset-specific operations (intersection, set difference, union, set restriction),

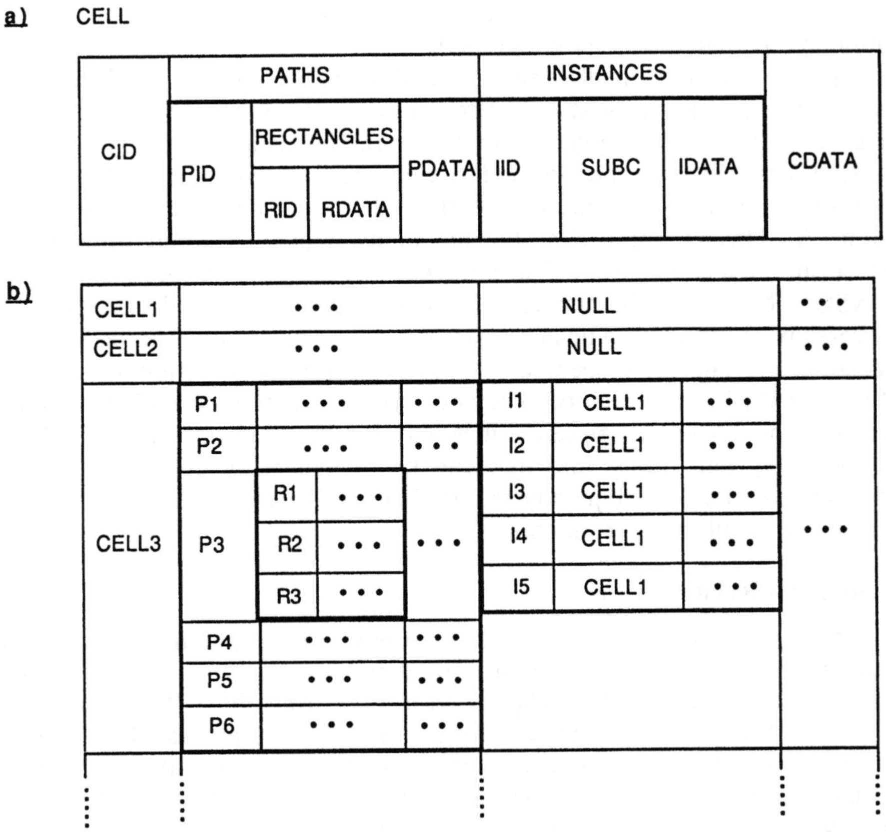

Fig. 2.80a,b. NF^2 database based on the hierarchical arrangement of Fig. 2.76. Composition is realized here by nesting. ISA-references are realized by CID entries under attribute SUBC. (**a**) Schema. (**b**) Example database

```
select *
from CELL
where CID = 'CELL3'

select P.RECTANGLES
from P in C.PATHS,
     C in CELL
where CID = 'CELL3'

select [CELLID : C.CID,
        INST   : (select I.SUBC, I.IDATA
                  from I in C.INSTANCES)]
from C in CELL
```

Fig. 2.81. Sample HDBL queries against the NF^2 database of Fig. 2.80: retrieve a complete tuple of the topmost relation (including all subrelations contained); retrieve a specific subrelation; construct a new (NF^2) table out of data retrieved from various levels of the table hierarchy

- conversion operations (ordering of multisets/lists, lists <-> multisets, 1Nf <-> NF^2 structures),
- duplicate suppression (in lists and multisets),
- counting list and multiset elements,
- arithmetic and related aggregation functions (sum, average, maximum, etc.),
- comparison operations for lists, multisets, etc., and
- constructors (for tuples, lists, and multisets).

Figure 2.81 gives some examples for the query language HDBL which extends SQL to cover the additional features of AIM. A cursor-based application programming interface is also provided.

Finally, AIM includes time version support. Historical states of tables or subtables may be retained instead of being replaced or updated. This enables *as of* queries that retrieve data reflecting the state at a specific time in the past.

Explicit Structuring Capabilities: DAMOKLES

DAMOKLES is a database system prototype currently developed as the basic information management component for a software engineering environment. As the specific requirements of this particular application area are not much different from other design applications, the DAMOKLES DODM data model belongs to the same class as the ones mentioned above.

However, DODM does not try to plant molecular objects into a basically relational framework. Instead, it provides an extended entity-relationship data model, which is then no longer only a semantic framework, but includes a full-fledged set of operators. The main structural features of DODM include the following:

- molecular entities (called *structured objects*) which may be built recursively and which may also overlap (i.e., a subentity is part of two or more superentities), so that network-like structures, not just hierarchies, are possible;
- versions of structured objects, supported by a basic mechanism like the one in XSQL, however, versions are relative to a structured object and do not introduce a new higher-level construct like a design object; and
- arbitrary n-ary relationships between objects of any granularity including versions; cardinalities are also supported.

Structured Objects

A *simple* DODM object, like records/tuples in classical data models, is composed of a number of *attributes*. One or more of these attributes may be designated as the object *key*, and thus required to be unique within the set of current database objects of the respective type at any point in time. In addition, DAMOKLES automatically assigns a surrogate (called OID) to any object upon its creation. Attributes form the *descriptive part* of an object. *Structured objects* also consist of a *structural part*: It includes a set of *subobjects* (perhaps with relationships among them) that, in turn, are objects in their own right and thus may themselves be simple or structured.

In the database schema, the descriptive part of an object type is specified as usual by enumerating the desired attribute names and their associated value sets. For the structural part (if any), the type names of desired subobjects are listed. As there are no further restrictions,

- recursive objects may occur by using (directly or indirectly) their own object type within their structure, and
- structured objects need not always be simple hierarchies of lesser objects but may overlap in arbitrary ways.

Instances of structured object types originate in two ways. One always starts from instances of the subobject type. In a kind of AUTOMATIC option, the creation of a subobject is accompanied with the creation of an instance of (one of) its superobject types. Alternatively, the subobject is manually inserted into its superobject at some later time. Subobject removal and automatic subobject deletion upon superobject deletion are also supported. Operators for objects further include:

- locating in sequential order the objects of a given type,
- retrieval based on object identifiers or on attribute values (using a cursor),
- individual or joint attribute retrieval and modifications,
- navigation within structural objects to the next subobject of a given type,
- locating the next structured object in which a given object participates (remember that structured objects may overlap), and
- copying object attributes or entire objects.

Object Versions

Object versions allow the representation of multiple instances of the (semantically) same object under the auspices of the DBMS. The main characteristics of the DODM concept are as follows:

- Versions are always associated with objects or, more precisely, each version belongs to exactly one object, its *generic object*.
- Both the generic object as a whole and its individual versions may have attributes and an internal structure. While the attributes and internal structure of the generic object are supposed to be common to all its versions, the version attributes and the version composition from subobjects may differ between versions.
- Generally, versions can be treated as objects in their own right. Linguistically, their type is denoted as <object type name>. VERSION with <object type name> is the name of its generic object.
- Consequently, all versions of an object have the same kind of structure and the same attributes. They may even have versions themselves. Both the entire object (with or without all its versions) and individual versions may be referenced.
- Among the versions of one object, an implicit predecessor-successor relationship is maintained which may optionally be linear, tree-like, or acyclic; versions are numbered in creation sequence.

- Operators on versions allow sequential location of versions in the version graph structure or by version number, location of the generic object of a given version, and insertion and removal of a version into or from the version graph of a generic object, respectively.

Relationships

Relationships are n-place ($n \geq 1$) bidirectional associations of objects. Each place is characterized by a *role attribute*, and relationships in their entirety may possess further attributes. Similar to objects, the database schema describes relationship types with corresponding instances in the database. As an inherent consistency constraint, the user may specify a minimum and maximum *cardinality* for each role. It defines how often an object must at least or may at most participate in a role of a given relationship type.

Relationships play a major part in defining structured objects: Like subobjects, they may be included in them. Obviously, objects at any level, generic objects, and individual versions may all be used to define relationships. Thus, all kinds of inter-object or intra-object associations may be defined, and a powerful set of operators (similar to those for objects) allows exploitation of them.

Discussion

In principle, the object-subobject structure of complex objects as well as object versions may be regarded as nothing but special cases of general relationships. There are at least two reasons that justify the dedicated concepts chosen for the DODM:

- Since complex objects and object versions are standard requirements for the application area of interest, it is more convenient for the database designer to have appropriate counterparts in the data model. Also, as experience shows, the schema becomes much more comprehensible than a purely entity-relationship schema.
- As the DBMS recognizes the special semantics of these concepts, it may provide efficient implementations for complex objects and versions (e.g., delta storage [Dadam et al. 84, Tichy85] in the latter case).

Let us now illustrate the DAMOKLES DODM concepts by some examples. Figure 2.82 shows the schema and a small database for the running VLSI example. In Fig. 2.83, a modified definition of the object type CELL is given that allows for the creation of CELL versions: While the CID attribute pertains to the generic object as a whole, the example assumes that CDATA and the set of (PATH-)subobjects are specific for every version. Finally, Fig. 2.84 gives an impression of the DML facilities. They are currently available as procedures that may be called from C programs.

2.4.2.4 Attributes

Complex objects, object versions and relationships express structural properties of real-world entities. Descriptive properties (of both objects and

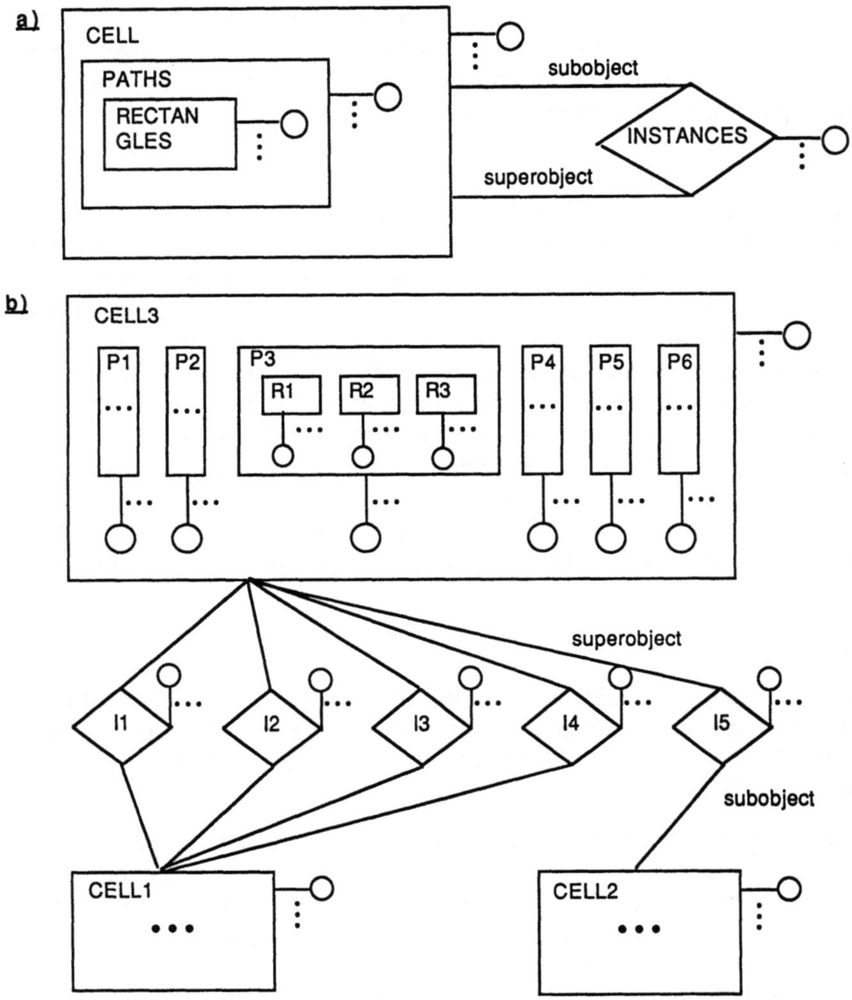

Fig. 2.82a,b. DAMOKLES DODM-example: schema (**a**) and example database (**b**) in graphical notation (circles represent attributes of entities/relationships)

relationships) are captured by attributes whose values are taken from predefined value sets called *domains*. Integers, reals, characters, and (fixed or variable length) texts are popular domains in both traditional and engineering database systems.

Some of the latter systems also include facilities to construct non-atomic attribute values, much like data declaration mechanisms in higher level programming languages. Type constructors for *vectors* or *matrices, lists,* and *(multi-)sets* can be found in one system or the other. As discussed above, surrogates and explicit or restricted references are also handled as special-case attribute types. Furthermore, several new approaches under research allow for the definition of *abstract data types* to support user-specific domains. In this

```
object type CELL
   attributes
      CID : ...
      CDATA : ...
   structure is PATHS
end CELL

object type PATHS
   attributes
      PID : ...
      PDATA : ...
   structure is RECTANGLES
end PATHS

object type RECTANGLES
   attributes
      RID : ...
      RDATA : ...
end RECTANGLES

relationship type INSTANCES
   relates
      PROTO : CELL
      SUBC  : CELL
   attributes
      IID : ...
      IDATA : ...
end INSTANCES
```

Fig. 2.82c. DAMOKLES DODM-example: textual from of the schema of Fig. 2.82a

```
object type CELL
   attributes
      CID : ...
   versions linear
      (attributes
         CDATA : ...
      structure is PATHS)
end CELL
```

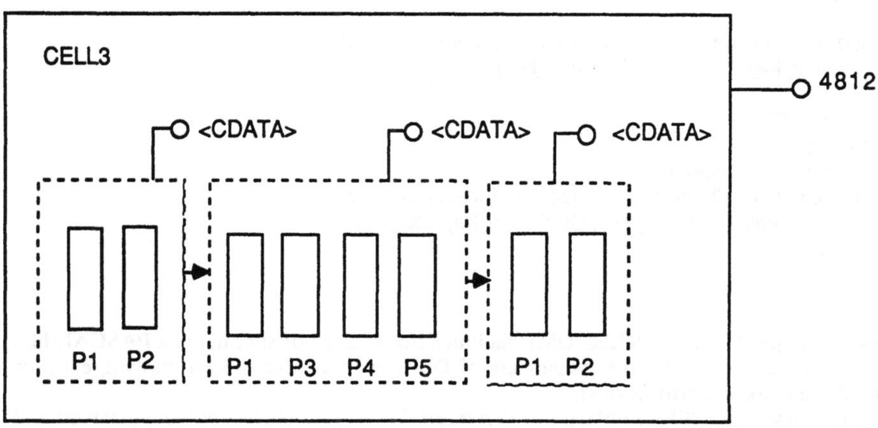

Fig. 2.83. CELL type with version; sample CELL object with three versions containing different sets of subobjects. Note that a linear sequence of the versions has been specified (tree-like or arbitrary acyclic arrangements would have been other choices)

```
1  cbuf.cdata := . . . ;
2  db_io (IN : cell, cbuf, OUT : c_key);

3  for i := 1 to 6 do
4     begin
5     pbuf.pdata := . . . ;
6     db_io (IN : paths, cell, c_key, pbuf, OUT : p_key[i])
7     end;
   .
   .
   .
8  for i := 1 to 3 do
9     begin
10    rbuf.rdata := . . . ;
11    db_io(IN : rectangles, paths, p_key[3], rbuf, OUT : r_key[i])
12    end;
   .
   .
   .
13 roles[1].ro_name := PROTO;
14 roles[1].ro_okey := c_key;
15 roles[2].ro_name := SUBC;
16 roles[2].ro_okey := cell1_key;

17 for i := 1 to 4 do
18    begin
19    ibuf.idata := . . . ;
20    db_ir (IN : instances, ibuf, roles, OUT : ikey[i])
21    end;

22 roles[2].ro_okey := cell2_key;
23 db_ir (IN : instances, ibuf, roles, OUT : ikey[5]);
   .
   .
   .
24 number := 0;
25 inst_key := nokey;

26 db_fioo (IN : db_first, cell, subc, instances, proto, cell, cell1_key,
27             inst_key, OUT : obj_key, inst_key);

28 while not EOT
29    do begin
30       number := number + 1;
31       db_fioo (IN : db_next, cell, subc, instances, proto, cell,
32                   cell1_key, inst_key, OUT : obj_key, inst_key)
33       end;
```

Fig. 2.84. Example of DAMOKLES DML facilities (for reasons of simplicity, a PASCAL-like syntax has been chosen and some parameters of DML operators have been omitted, e.g., for database designators and error codes).
lines: (1)–(2): insert the CELL3 object; (3)–(7): insert the PATH3 objects into the database and make them subobjects of CELL3; (8)–(12): analogously, insert RECTANGLES objects; (13)–(23): insert the INSTANCES relationship objects as derived from Fig. 2.82; (24)–(33): determine the number of instances of CELL1 (how often is it used as a subcell?)

case, values are usually represented by a number of bytes, and specific operations have to be programmed which are then the only way to read and manipulate values. For example, an abstract attribute type "box" with operations to move and rotate boxes (in a system of coordinates) and to calculate box sizes might be defined based on a byte string that comprises two pairs of x-y coordinates for the lower left and the upper right vertices.

Finally, *long fields* are provided for data whose internal structure is not to be exploited by the database system (because this would be too costly or because the structure is not relevant for the intended application, e.g., a textual description of a VLSI cell). Long fields behave similarly to direct files: They are accessed by specifying a starting position and a length that indicates the number of bytes to be written/read/modified at this position. Note that by means of long fields, traditional files can be incorporated into database systems. The benefits are that it is unnecessary to deal with two systems simultaneously, and that they can share recovery, access control, and other mechanisms.

2.4.2.5 Object-Oriented Database Systems

Novel data models as presented above fall into a class that is currently attracting more and more interest from both researchers and practitioners: **object-oriented data models** (database systems based on such models are called object-oriented database systems).

The general approach is to provide facilities to define and deal with information units that allow direct representation of real world entities of any structure and complexity with respect to subunits and attributes, and of any operability. Depending on the nature and degree of object-orientation they provide, the following classes of approaches can be distinguished:

- Structurally object-oriented models support complexly structured objects , i.e., objects that may be constructed from subobjects (which are objects in their own right) and thus go beyond incorporating simple attributes of the usual types. Structural object-orientation also includes a set of appropriate operators to deal with objects. These operators are **generic** in the sense that they apply to all object types that can be defined using structural constructor mechanisms of the data model; they are thus not type-specific and are predefined in the data model itself. All of the approaches discussed in Sect. 2.4.2.3 were of this kind.
- Behaviorally object-oriented models provide facilities that allow the definition of new object types **including** the appropriate type-specific operators. In the simplest case, these types act as attribute domains only, and their values are represented as byte strings that are interpreted by the operation code that the user has to produce as part of the database design. Behavioral object-orientation has been influenced by abstract data type concepts in the programming language domain.
- Fully object-oriented models combine structural and behavioral object-orientation: Complexly structured objects may include instances of user-defined types, and user-defined types may use complex object structures as their representation.

Primarily for behavioral objects, **inheritance** mechanisms are offered that allow establishment of a taxonomy among object types: Objects of a type which is considered to be a specialization of some other type (subtype or supertype) inherit the properties (operations and attributes) defined for the supertype; they may have further properties of their own (is-a property, see Sect. 2.4.2.1).

As mentioned above, the approaches presented earlier fall into the class of structural object-orientation which is both appropriate for many engineering applications and sufficiently understood that systems can be expected in the marketplace during the next couple of years. For behavioral object-orientation, some start-up companies already offer first products which are mostly based on the Smalltalk programming language, and add object persistence and other database features to it. However, they do not explicitly support complex object structures as yet. This would only be true for fully object-oriented systems which are, however, still in their infancy these days.

2.4.3 Other Issues

As we have mentioned, issues beyond the data model such as interfaces, consistency, concurrency, recovery, and transaction management are not well advanced yet for new approaches to engineering database systems. Though numerous researchers are active in this field, no clear directions can be identified to date. We will therefore only summarize these issues in this chapter.

With regard to **interfaces**, it is clear that conventional query languages are no longer sufficient to cope with the extended functionality of the new data models. Where complex objects etc. are introduced by extending the relational model, the query language is usually SQL with appropriate extensions (as, e.g., for NF^2). In the case of XSQL, a hybrid approach has been taken: The relational view of the database can be exploited by pure SQL, and the tuple-by-tuple construction of objects uses SQL as well. For fetching, copying, etc., of complex objects, and for using the version concept, however, procedural commands have been provided. DAMOKLES currently implements a completely procedural interface. Studies have been done on an SQL-like language that offers facilities for advanced data model concepts and is straight SQL if only flat objects without versions, etc., have to be dealt with.

Note that interactive interfaces are not a first priority issue for most engineering database applications: Database access is primarily through tools, and only to a fairly small extent will human users query the database directly. Of course, this is no argument against descriptive (in contrast to procedural) languages which are generally considered to provide more convenience for programmers and interactive users.

Consistency in engineering databases encompasses the need for numerous and often complexly structured constraints, for flexibly definable checking times and reactions to constraint violations. As this is far from what traditional mechanisms can offer with reasonable efficiency (and especially without penalizing applications that do not need it), it is often argued that most of consistency control should be maintained in the application programs as it is right now. However, this attitude clearly leads to less integration and quality of data management than is desirable, and thus it is only acceptable as an implementation as long as no appropriate database mechanisms exist.

Indeed, there are some approaches currently in use that overcome the detected shortcomings. One of the most advanced ones is a so-called Event/-Trigger Mechanism that allows for the definition of events that can be raised at any point in time during database processing (i.e., not only at the end of work units or at the end of specific database operations). Actions (i.e., arbitrary programs which may or may not access the database) may be associated with events by means of triggers: The effect is that once an event is raised, the associated actions are executed before normal processing is resumed. With respect to consistency control, the Event/Trigger Mechanism is used with checking routines as the actions. These may either be introduced directly, or they may be generated from descriptive formulations of consistency rules in an appropriate language. At the end of checking, an event signaling either failure or success is raised, and triggers to respond to, e.g., a constraint violation may be defined as desired.

In summary, the Event/Trigger Mechanism forms a powerful basis on which consistency control for systems with demanding requirements as mentioned above may be built. Still, it is not at all easy to provide powerful constraint definition languages and, especially, to determine all the constraints that need to be captured for proper design data management.

Of course, a data model in itself also incorporates a small number of fairly straightforward (but nevertheless important) consistency rules. Traditional features like type consistency (e.g., according to the schema definition, only values of specific types are accepted for a given attribute), or uniqueness of defined keys are provided in the previously mentioned data models as well. Moreover, e.g., DAMOKLES allows for the definition of role cardinalities in relationships that are later observed by the database system. It also supports referential integrity (the deletion of an object causes the automatic deletion of all relationship instances in which it participates).

Concerning **concurrency control** and **transaction management**, engineering database systems have to take into account the length of meaningful work units. Thus the conventional transaction concept has to be augmented by approaches that are currently discussed under names like long-lived transactions, conversational transactions, or nested transactions. Again, there is no concept yet that is generally understood and accepted as in the traditional case. The lines along which prototype mechanisms in this area work are as follows:

- Every user works in his own private workspace or database; data objects which should be generally available are placed in public databases (of varying accessability).
- For long-term processing of an object, the user has to check it out to his private database; after termination, he checks it in.

- During check-out periods, locks are set for objects in the public databases. However, lock modes can usually be handled less rigorously than for short transactions: In some cases, concurrent read and write accesses may be tolerated, and only two write locks conflict. Of course, more sophisticated approaches may be devised.
- Instead of putting a user's long transaction into a wait state if a desired object is not available due to lock conflicts, the request is denied and the user can again decide how he would like to continue his work. He may be notified about who is currently holding the object in question and can then "manually" negotiate with the holder (note that this protocol is pretty much the same as with real life "checked out" items, e.g., books in a library). In the presence of a version concept, one may also define that in case of write conflicts one simply creates a new version and decides afterwards as to how to go on.

Long transactions also require specific **recovery** precautions as a rollback of the entire unit may delete too much meaningful work. This is achieved by either considering a long transaction as a kind of umbrella around a sequence of short transactions that are merely used for recovery purposes (and not for consistency control or synchronization; so-called recovery transactions), or by providing a special feature to set and potentially reset to savepoints within a long transaction.

Finally, transaction and recovery concepts should take into account the typical hardware architecture used for many engineering database systems, namely a network environment. The set of private and public databases lends itself for distribution across the nodes. Check-out/check-in thus usually also means bringing objects to the user's workstation, where many of the recovery precautions should be taken (while traditional ones will be applied at the server site).

A number of sophisticated proposals in this area have recently been published that try to combine all issues mentioned and several more. It is not at all clear, however, whether any of these will gain widespread momentum in the future.

2.4.4 Further Reading

As the topics discussed in this section are all comparatively new and unsettled, there is practically no textbook-style literature available to date, and the wealth of papers found in journals and conference proceedings is often difficult to read for practitioners. Let us nevertheless mention a few references that might help the interested reader find his way to more details.

The following papers have been mentioned in the text and cover some of the material presented in more depth:

Dittrich K.R., Lorie R.A.: *Version Support for Engineering Database Systems.* IEEE Transactions on Software Engineering 14 (1988), 425-437

Dadam P., Lum V., Werner H.-D.: *Integration of Time Versions into Versions into a Relational Database System.* Proc. 10th Internatl. Conf. VLDB, 1984, 509-522

Tichy W.F.: *RCS – a System for Version Control.* Software Practice and Experience 15(1985)7, 637-654

Other publications on the data model approaches discussed include:

Lorie R., Plouffe W.: *Complex Objects and their Use in Design Transactions.* Proc. SIGMOD Database Week/Engineering Design Applications, 1983, 115-121 (on XSQL)

Dadam P., et al.: *A DBMS Prototype to Support NF^2* Relations: An Integrated View on Flat Tables and Hierarchies. Proc. SIGMOD Int. Conf. on Management of Data 1986, 356-367 (on AIM)

Dittrich K.R., Gotthard W., Lockemann P.C.: *DAMOKLES – a Database System for Software Engineering Environments.* Lecture Notes Comp. Science 244. Springer, 1987, 353-371 (on DAMOKLES)

An introduction to object-oriented database systems in general is given in

Dittrich K.R.: *Object-Oriented Database Systems.* in: Spaccapietra, S. (ed.): Entity-Relationship Approach. North-Holland, 1987, 51-66,

and the following proceedings of a workshop and additional papers show the variety of topics that are currently discussed under this headline:

Dittrich K.R., Dayal U., (eds.): *Proceedings 1986 International Workshop on Object-Oriented Database Systems.* IEEE Computer Society Press, 1986

Atwood T.A.: *An Object-Oriented DBMS for Design Support Applications.* Proc. COMPINT 1985, 299-307

Some of the non-data-model issues are treated in:

Dittrich K.R., Kotz A.M., Mülle J.A.: *An Event/Trigger Mechanism to Enforce Complex Consistency Constraints in Design Databases.* SIGMOD Record 15(1986)3, 22-36

Kim W., et al.: *A Transaction Mechanism for Engineering Design Databases.* Proc. 10th Internatl. Conf. VLDB, 1984, 355-362

Finally, we mention three references that might be of general interest for the reader of this chapter:

Katz R.H.: *Information Management for Engineering Design.* Springer, 1985 (though in book form, it is an edited collection of individual papers)

Kemper A., Wallrath M.: *An Analysis of Geometric Modeling in Database Systems.* ACM Computing Surveys 19(1987)3, 47-91

Stonebreaker M., Rowe L.A.: *The Design of POSTGRES.* Proc. SIGMOD Int. Conf. on Management of Data 1986, 340-355

3 Utilization of Engineering Databases

3.1 Motivation

The difference between requirements of engineering databases and those of more traditional applications together with the different database concepts to satisfy them have already been discussed in detail in Chap. 2. As we have seen from Sect. 2.4, a variety of proposals for their solution exist in modern database research, but no systems are as yet available in the marketplace.

Current developments in the wide area of engineering databases have their roots in different research directions: on the one hand, engineers in CAD/CAM and graphics-oriented applications, and on the other hand, database researchers. Future commercial products should aim at the integration of concepts from both sides. The acceptance of such systems by users depends mostly on the database application interface and system support for modeling and manipulating the applications data.

The intent of this chapter is to present examples of such engineering database interfaces and how to work with them. We do not introduce new concepts but rather use those of Chap. 2. The operations and resources available will be shown in examples, and some shortcomings of those concepts from an application point of view will be pointed out.

3.2 Database Schema Design

Before user data may be inserted or manipulated with manipulation operations (DML), a database schema has to be defined for the database by an authorized person (e.g., a data designer in cooperation with the project administrator).

The first part of Sect. 3.2 shows some principles of data definition. Some typical situations when the applications data are organized for mapping into the database are described in Sect. 3.2.2. Objects can be classified by different characteristics, which can not necessarily be represented by certain object and relationship types. These description criteria are explained in Sect. 3.2.3.

In Sect. 2.4, the DAMOKLES data model was introduced as an example of an engineering database system. The usage of the features of the data model should show how to define data structures and data types for a specific application. Therefore we first present an example from hardware design, which is

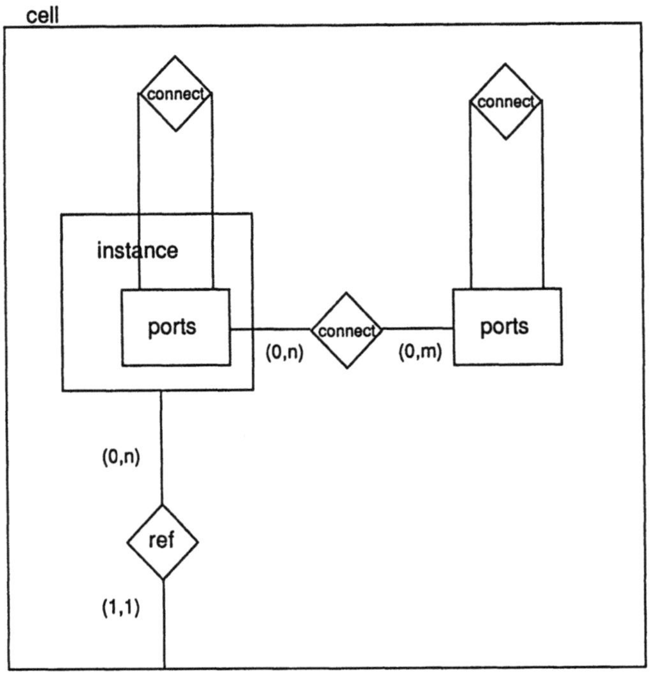

Fig. 3.1. Cell example as ERM diagram in DAMOKLES

used to provide some basic ideas for schema definition in the following sections. The example object is shown in Fig. 3.1. Pictures or diagrams help to define a database schema; for relational databases, tables show the structure of the database; and for extended Entity Relationship Models (ERM), we use the symbols of ERM diagrams to identify the objects and relationships in the data model's conventions.

The first step in schema design is the identification of objects and their structure for the description of cell compositions. For the example of VLSI cells we have to model cell prototypes, and instances of cells and ports. Between them exist certain relationships or dependencies. With DAMOKLES we have features like objects, structured objects, and relationships. The schema designer has to think in types of objects or relationships, because he defines classes of allowed objects and structures later for application programs or interactive database users.

3.2.1 Principles of Schema Definition

The administrator or schema designer has to create a model for the description of a given application. From this point of view the application is named the miniworld, because only a very specific part of the real world will be described by a model. Schema definition is modeling the user's miniworld semantic, and features like the data definition language (DDL) and support tools assist users to exploit the data model. A future trend in this direction is the use of graphical tools or interfaces for definition of a database schema. Systems support the assumption that human beings think in pictures, and therefore prefer to use diagrams or similar techniques.

The process of modeling design information with logical database structures is intuitive and varies from application to application. Although we can not give a recipe for defining schemas, we can formulate a number of requirements that are especially suitable in engineering applications.

Schema definition is the most important part of the integration of tools/-applications and database systems, because the schema is the cardinal point of the interface between both components. Data modeling is the process of formal description for structuring and describing user data by defining:

- object types,
- relationship types, and
- consistency rules (e.g., object control procedures).

In the following we concentrate on determining object (entity) and relationship types as the main parts of the data structuring process. The engineer as data designer has to attend to some principles.

When describing the schema, he defines the *data types* (i.e., object and relationship types) of interest. A database adhering to a given schema contains objects and relationships (*instances*) of the types defined in the schema. Object and relationship instances (and not types) are inserted and manipulated by entering and changing values. In discussing object semantics, this distinction is often forgotten. It is, however, important when talking about things like types, objects, configurations, versions, and so on (see also Fig. 3.2).

We can compare these two parts with computer programming languages; the data definition is similar to the declaration part, and the data manipulations are the program statements. However, the schema definition might not be as static as the type declaration in programming languages.

Design is an *evolutionary process*; that means object and relationship types evolve during the design phases. When declaring types in the first place, the data designer has to keep in mind that they must be *extendable*.

When modeling fixed *relationships* between object types like, e.g., the *resulting framework*. Consider a data model which allows a subobject to be part of more than one object at the same time, as in Fig. 3.3.

The careful use of link types leads us to the problem of *redundancy*. Sometimes structuring of data is surveyable, but there is a lot of redundancy in the information, because the features of the data model are not efficiently used. By structuring the data we can avoid some redundancy, if we have a global overview of the overall schema and the whole set of design objects. Redundancy is also a problem of consistency. Multiple copies of the same object lead to inconsistent states in the database, if one of them is updated.

Finally, the user's intention is to have not only a good logical representation of his data, but also efficient access to them. In current data models one can not avoid doing some physical design, which can be the definition of access paths or other physical organization of data. So we should not forget this point in the schema declaration either. For example, if the model supports

```
OBJECT TYPE  cell
        ATTRIBUTES
        c_name : name;
        released : boolean;
        author : name;
        . . .
        STRUCTURE IS
        ports,instance,ref
        AT MOST ONCE (ref)
        AT LEAST ONCE (ref)
END cell;

RELSHIP TYPE ref
        RELATES
        instance,cell
        ATTRIBUTES
        . . . .
END ref;
```

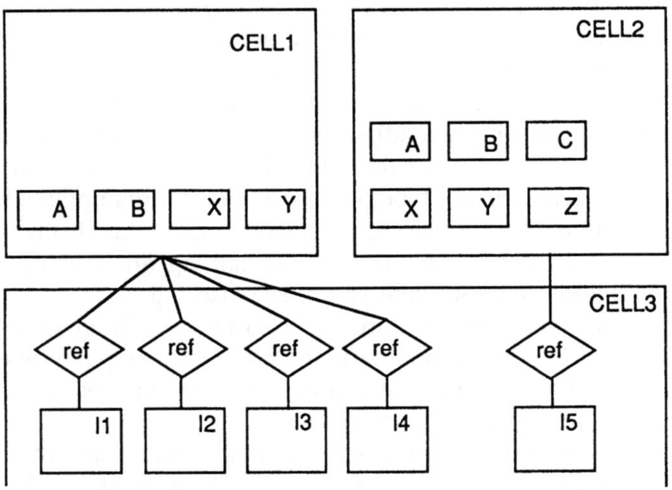

Fig. 3.2. Types and instances of objects

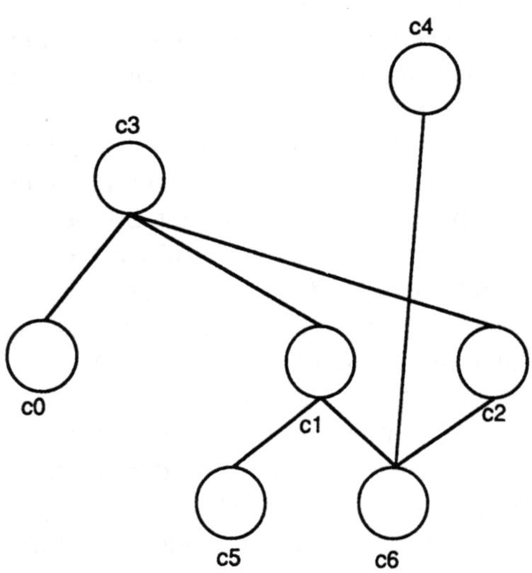

Fig. 3.3. Shared subobjects

the distinction between simple and structured objects, we should only declare a simple object type for unstructured simple data values. Otherwise, the overhead at the instance level, e.g., for retrieval, is too high. The conclusion is to be careful in using complex features of the model for simple solutions, because every service costs a price.

3.2.2 Typical Situations

Basic data structures as supplied by the data model (e.g., object and relationship types in DAMOKLES) are used for the logical organization of data. To define a certain schema (for a specific application), the designer has to group and arrange his data types in a way which, on the one hand, allows the tool or the user to insert and manipulate data in a simple way. On the other hand, the designer should take advantage of the whole expressive power of the data model. Only in this way is an optimal use of the database and its DML operations possible.

Consider the information about cell ports in a VLSI design. In the schema description one can fix the types of ports, or this information can be variable. Later in the design phase, the application has more flexibility if the type is not fixed (see Fig. 3.4).

In definition D1 the attribute type for *ports* is not defined exactly, and it is now the task of the applications to fix and interpret the semantics of a special port. In this case, the engineering database system can not support control of cell ports. Defintion D2 restricts the values allowed to be inserted as *port* attributes, and establishes a well-defined environment for manipulation.

```
OBJECT TYPE     cell
                ATTTRIBUTES
                Id_Nr: INTEGER
                generation: DATE
                Simulated: BOOL
                . . .
                Ports: STRING (10)
 . . .
 END

OBJECT TYPE     cell2
                ATTRIBUTES
                Id_Nr: INTEGER
                generation: DATE
                Simulated: BOOL
                . . .
                Ports: ENUM (IN,OUT)
 . . .
 END
```

Fig. 3.4. Fixing semantics by attribute domains

Database designers have the very difficult task of fixing as much semantics as possible. If a great deal of semantics is fixed by the model, the user has a well-defined environment and it is guaranteed that special tools will work in the right way. If less is fixed in advance, later users have more freedom in writing their own application programs.

Therefore, in an engineering design system, where user written DML-programs are not the primary application, but tools working on data, the administrator should fix as much semantics as possible. Of course, there are some counterexamples, but in an integrated system which should support and control very different tools, it is necessary to fix much semantics.

Schema design first requires an information analysis of the existing application specific tool environment. Which information flows between which tools, i.e., what kind of data are the input and what kind of data are the output of a certain tool or application?

After getting this information and structuring the data, one can begin with the data description in the database. The administrator has to model the representational and structural details of a design and the design process. There are a number of requirements for the database interface, which ensue from the structuring process and the produced data.

Engineers work with interactive graphic tools at their own workstation. After having reached a certain design stage, they apply some tools to check the design or to simulate and test it. Therefore, the whole design exists in different representations or views on different abstraction levels with, additionally, some interrelationships across the representations. Moreover, it must be possible to partition the design among subsystem designers. The database must provide structures for organizing the design description within and across representations, design versions, and interface specifications.

3.2.3 Objects and Interdependencies

There are many different characteristics that together determine the semantics of objects, which can be divided into three classes:

- object representations,
- object interdependencies, and
- library or object-overlapping information.

In the following subsections we consider these three main characteristics in more detail. From each of the following points, we can derive some requirements for the data structures and relationships needed in the design description of the database.

3.2.3.1 Object Representations

Representations in the context of design objects mean the description of an object at different abstraction levels. A complex object exists in more than one representation. Some kinds of representations are not application-dependent, and we can find some common points and give some examples. Objects can be described by:

- internal and external information,
- interface information,
- behaviour,
- documentation, and
- graphical information.

The schema should include representations of *internal and external information* belonging to an object. In certain levels, objects may be handled as black boxes with a special input and output. If the designer wants a differentiation for a cell component between, e.g., the name, colour, and dimensions as external and the number of primitives, geometry of primitives, etc., as internal information, he has to define two representations of the object. The distinction is significant: If a user wants to place this object in a certain drawing environment, then he only needs the external information. Another example is a program module: Its name, together with input and output parameters, may be defined as external, while the subroutine bodies or other details are internal hidden information.

Some objects have interfaces to other objects in the database, e.g., if an object is used within another one. Therefore the database has to manage *interface information* of an object, too. Consider once again the program module example. Interface information may be the function of the program, the programming language, and a reference to its external information.

Design objects need a *behavioural description*, if they are functional parts of a system configuration. There are many different forms for the behavioural description of design objects. It may be state diagrams and tables, informal text, logical truth tables, or other mathematical descriptions, etc. The formal descriptions are mostly on a high logical level and so are the features of a data model. So we can find a mapping between both forms.

Objects are often very complex and not self-descriptive. Therefore, they need a textual *documentation*; it should be cast into long fields or a so-called internal contents of database objects, without needing to be partitioned in an impractical way (see also the type definition in Fig. 3.5 for an example of mapping text into objects).

```
OBJECT TYPE     documentation
                ATTRIBUTES
                Id_Nr: INTEGER
                Creation: DATE
                Text: LONG_FIELD

                UNIQUE Id_Nr

...
END
```

Fig. 3.5. Defining text

Engineering design often produces objects which have a *graphical representation*. Graphical information like geometric parameters of an object is highly structured, and consequently it must be mapped into the appropriate data model features. Note that some graphical information might also appear in the internal or external parts mentioned above.

One point of dicussion when organizing a database for VLSI design with cells is the construction of a composed cell in different representations. We have to decide whether the composition of one cell is the same in each representation or not. Consider five different representations like functional description, logical description, behaviour or logical simulation, symbolic layout, and geometric layout. The VLSI cells are completely described if there exist correct data for each abstraction level, and correct relationships between these representations. What correct means depends on the schema definition and integrity rules defined by the application.

With different compositions of one cell for each representation, the description of dependencies between representations of one cell on one side, and compositions of cells by other cells in different representations on the other side, will be more complex. If representations relate to the same composition for one cell, addition of special information within representations is necessary if the composition deviates from a certain representation.

In the example, the composition of cells is uniform for all representations, and variations of cell compositions will be modeled by versions of cells. Each cell or version of cell can be related to special representation descriptions as in Fig. 3.6. We concentrate on the composition, which is now inside the cell node of the above figure. Here nodes represent structured objects and links are special relationships between them.

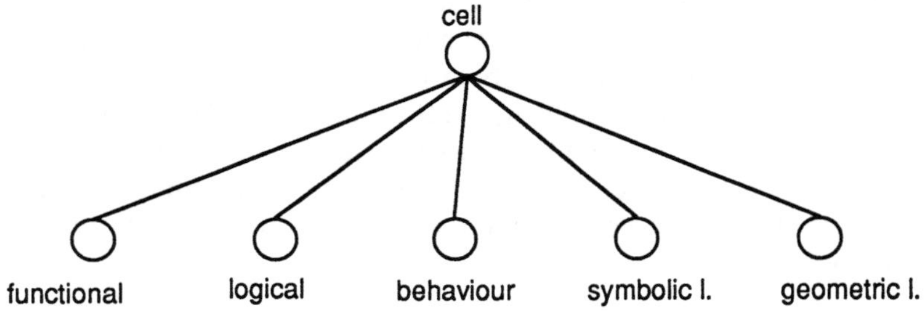

Fig. 3.6. Cells and representations

3.2.3.2 Object Interdependencies

An example of an object interdependency is the following: If object A is modified, the event induces an action (e.g., that object B must be regenerated). Such complex consistency constraints can not be enforced by the system in every case. Sometimes consistency is best maintained semi-automatically by sending messages to the user, or by prohibiting some operations on such data. Another possibility is a strict control over the execution of tools in a well defined order (e.g., to run a logic test before the physical layout of a circuit is done). However, there are some general interdependencies that the engineering database system can support like:

- composition,
- hierachy,
- connectivity information, and
- evolution.

The designer defines structures that express how complex objects are *composed* from primitives (basic objects and structures). We can describe a composed object as a set of independently manipulated primitives or complex objects. Between these objects, a characteristic relationship (e.g., a hierarchy or a network) exists. An example of a composed complex object is already given in Fig. 3.1. The complete type definition of this object will be shown in Fig. 3.7.

Hierarchical construction of the design is a result of natural top-down partitioning during the design process. The *hierarchy* is thereby a special relationship between objects, and the construction is a special method. Hierarchically structured systems are ubiquitous in most miniworlds. Among them are the 'component_of', 'child_parent', and 'is_a' relationships (partition/composition, specialization/generalization).

Connectivity information between objects which may be connected by certain relationships is difficult to represent. The problem is deciding where to declare it: as part of one of the objects, distributed, or as an object in its own right? You have to decide which object this information primarily belongs, and which kind of information should be "visible", if an object is touched. Another possibility for modeling this information is to define a set of attributes of the relationship type for relations which connect objects.

To model the *evolution* of design objects and interdependencies between them, the user should classify his objects as versions and use the system's version management. Some version mechanisms organize versions of an object in a graph, which represents version interdependencies in a sequential or tree-like way. Version management will be discussed in detail in section 3.3.

The main difficulty in abstracting from real world objects to database object types is the recursive definition of cell prototypes. Cells reference cells by definition of instances as parts or subobjects of cells, which stand in relationship (*ref*) to other cells. The example shows the difficult distinction between the type level and manipulation level with sample objects (see also Fig. 3.2). Notice that recursion with types is not the same as recursive objects!

Instances, Ports and *Ref* are parts or subobjects of *Cells* because their existence depends on the existence of a certain cell prototype. Figure 3.8a shows the DAMOKLES DDL for a VLSI cell schema.

Now we relate the cell composition to a certain representation, the logical description mentioned above. Then we have to define another structured object and a relationship *has_rep* as declared in Fig. 3.8b.

```
SCHEMA vlsi

CONST length = 30;

VALUE_SET

date : STRUCT
        day : INT SUBR [1..31];
        month : INT SUBR [1..12];
        year : INT SUBR [1986..2000]

        END;

name : STRING [length];

OBJECT TYPE ports
        ATTRIBUTES
        p_name : name ;
        i/o : direction ;
        . . .
END ports;

OBJECT TYPE instances
        ATTTRIBUTES
        i_name : name ;
        creation_date : date ;
        . . .
        STRUCTURE IS
         ports
END instances ;

OBJECT TYPE cell
        ATTRIBUTES
        c_name : name ;
        released : boolean ;
        author : name ;
        . . .
        STRUCTURE IS
        ports, instance, ref, connect
        AT MOST ONCE (ref)
        AT LEAST ONCE (ref)
END cell ;

RELSHIP TYPE ref
        RELATES
        instance, cell
        ATTRIBUTES          . . .
END ref ;

RELSHIP TYPE connect
        RELATES
        port, port
        ATTRIBUTES          . . .
END connect ;
```

Fig. 3.7. Type definition (DDL) of example Fig. 3.1

```
OBJECT TYPE logical_rep
        ATTRIBUTES
        l_rep_name : name ;
        text : longfield ;
        . . .
        STRUCTURE IS
        picture, gates, description
END logical_rep ;

RELSHIP TYPE has_rep
        RELATES
        logical _rep, cell
        ATTRIBUTES
        . . .
END has_rep ;
```

Fig. 3.8. Extension of type declarations

3.2.3.3 Library and Object Overlapping Information

Design objects are usually developed within projects. For each such project, global information is held to manage the different design parts, which arise during the phases of the system life cycle. Library objects or relations between objects and libraries are therefore important for design environments. We have to consider object overlapping information, which is centrally held in a library, including:

- project dependency,
- simulation information,
- technology information, and
- standard libraries.

An important piece of information for partitioning a collection of designs is the dependency between *projects* and objects. It is common practice that for each project, the adminstrator defines a new database because of organizational reasons. It should be possible to interchange project overlapping information or technical details for further design constructions between the different databases.

Simulation tools inspect existing objects and insert new ones. To simulate, e.g., the electrical representation of a VLSI cell represented as a complex object in the database, the designer has to map the features and semantics of a certain hardware description language into the data structures of the basis data model.

A design object in a certain representation may need some *technology* parameters to include the object in a final system configuration. Technology information is not usually private individual data, but rather it is, e.g., dependent on a workstation or dependent on a certain material. So the designer should define this information in a general library, accessible to all users who work on the same design in related projects. To use the technology data, one has to reference the related objects, or make entries in a technology-description table or entity.

Design data and structured objects are often reused in other designs. After completion of object preparation some objects will be released from private use to the global environment. Released objects will be compared with or included in the current development. Another point is the inclusion of standard solutions, e.g., VLSI cell design with standard cells is a common technique. In these cases we need *standard libraries* in a global database area.

What we have learned from our example is the transition from an abstract model of a special miniworld to a concrete description. We have to formalize the process of definition; this is much easier when regarding special features of a certain data model which supports engineering requirements, concerning mainly complex objects.

We suggest that the reader's next step should be to define and experiment with an example from his own application environment.

3.3 Version Management

Generally speaking, all technical design processes produce versions of the objects to be designed. It is impossible to give a formal specification of a version; versions may represent provisional results, erroneous developments, improved designs, updates, or final results. In particular, versions reflect special interrelations resulting from the design process. An idea of version generation is illustrated in Fig. 3.9.

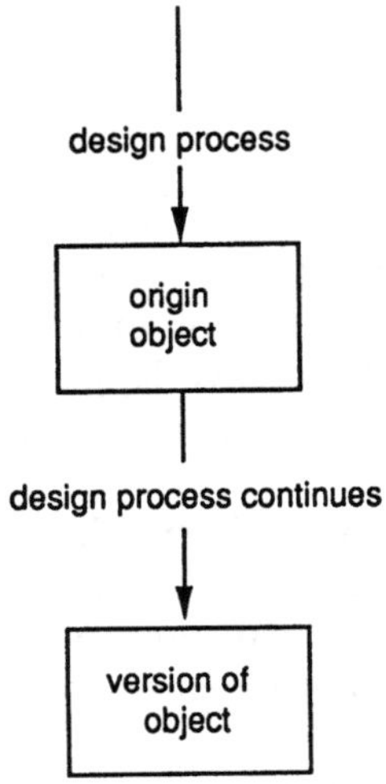

Fig. 3.9. Version generation process

The lack of a formal specification makes it impossible to determine the semantics of versions by a database system. The semantics of versions depend on the particular design process. Therefore, the support of version management by an engineering database system must remain restricted to some basic versioning mechanisms. A complete model of version interrelations is the responsibility of the design process, but should be founded on these basic mechanisms.

3.3.1 Version Generation in the Course of a Design Process

Each design process can be subdivided into different design phases. The sum of these design phases and their interdependencies sets up a so-called design life cycle. Remember our example of a VLSI design. In the case of a VLSI design, we can find at least four design phases:

- requirement engineering,
- conceptual design,
- logical design, and
- layout or physical design.

The first step of a VLSI design is to establish a list of design goals. The conceptual design results in a decomposition of the circuit to be designed into

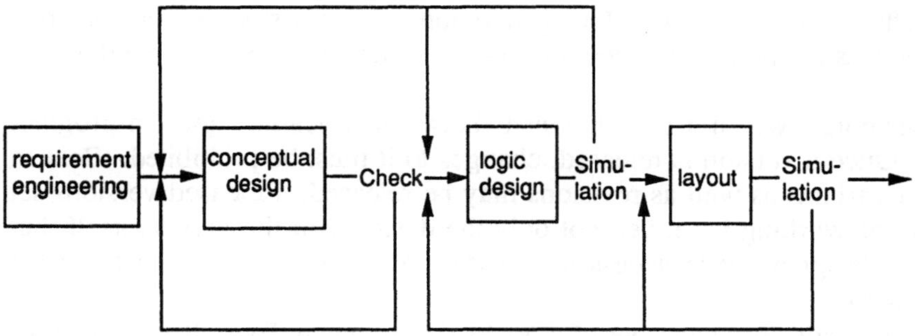

Fig. 3.10. Design life cycle

subsystems, and a specification of the behaviour and interactions of these subsystems. This way, the functional dimension of the circuit is given, e.g., by a register transfer description or block diagrams. The logic design leads to a logic description of the subsystems, usually presented as logic gate schematics.The last step in the course of VLSI design is to work out the layout. The layout describes precisely how to place and interconnect the design parts. Each design phase leads to a complete description of the circuit, but reflects a different level of abstraction – so-called representations of an object.

Of course a design process is not as linear as implied by the above explanation. As illustrated in Fig. 3.10, a lot of iterations may occur during a design process.

Each design phase is followed by a check for correctness. In our example of a VLSI design this can often be done automatically, e.g., the layout is checked by design rule checkers and the logic design by logic simulators, switch level simulators, or circuit simulators.

It stands to reason that detection of errors leads to iterations, i.e., the corresponding design phase has to be passed through again. Iterations are not only caused by error detection, but also by optimization of a certain design step.

Even if the results of a certain design phase are correct and optimized, redesign may be necessary. For example, the logic design does not take into account geometrical aspects. When designing the layout, there are geometrical restrictions we have to take into consideration. If it is not possible to design an adequate layout of our logic design because of such restrictions, we have to look for alternative solutions.

As shown above, many things can force iteration cycles, each one leading to a version of an object. This is not only a characteristic of VLSI design, but is typical of all technical design processes. We can find similar design phases in more or less all technical design processes.

Reflecting on our analysis, we may now distinguish between two classes of versions – variants and revisions. Variants of an object have similar functionality but may have different performance characteristics. For example, two layouts of a circuit may be designed, one consuming less area and another consuming less power. Variants reflect design decisions while revisions of an object reflect improvements. The generation of a revision is always done based upon its temporary predecessor, while variants may be generated at any time.

Furthermore, we must distinguish between released and in-progress versions. Once a version is released, changes to it must be prohibited. Bear in mind that variants as well as revisions may be released. Released versions set up a kind of working base, i.e., not only the designer of the version itself, but also other designers have access to released versions and may use them for ongoing work.

Let us consider our example of a VLSI design again. Even if the layout of our circuit is not optimized, a prototype based on it may be produced. In such a case we have to guarantee that the version of this layout will neither be changed nor deleted. In contrast to released versions, in-progress versions are the ones still under development. Hence, a designer should be able to change, improve or delete them.

To summarize, we should bear in mind that each design object is described on different levels of abstraction (*representations*). Usually various versions of the designed objects exist for each of these representations. There are two classes of versions to be managed during the design process: revisions and variants.

Of course, versions may have further meanings. In the present context it should suffice to focus on the classes given above. Other semantics of versions are not yet supported by engineering database systems .

3.3.2 Modeling of Version Interrelations

Version interrelations do not only exist within a certain representation, but also across different representations. Modeling of version interrelations across different representations is a very difficult task. Therefore we must focus on the modeling of version interrelations within a certain representation.

To manage version interrelations, we need a modeling scheme. A modeling scheme depends on what interrelations a designer is interested in. Thus we have to work out different schemes depending on the interrelations to be managed.

Often version management is needed to manage history information only. In this case a linear modeling scheme is sufficient, even if it implies sequential version generation as illustrated by the following example (Fig. 3.11).

In this example, we assume that v_0 is the generic version of the object to be designed and the numbering of versions reflects their order of creation. Then our scheme implies that version v_1 is a revision of version v_0, version v_2 a revision of version v_1, etc.; i.e., each version is based on its chronological predecessor. However, such a temporal sequence need not reflect the real design interrelations. For example, version v_2 may be a redesign of version v_0 because version v_1 turned out to be a dead end. Hence a linear modeling scheme does not allow distinction between variants and revisions.

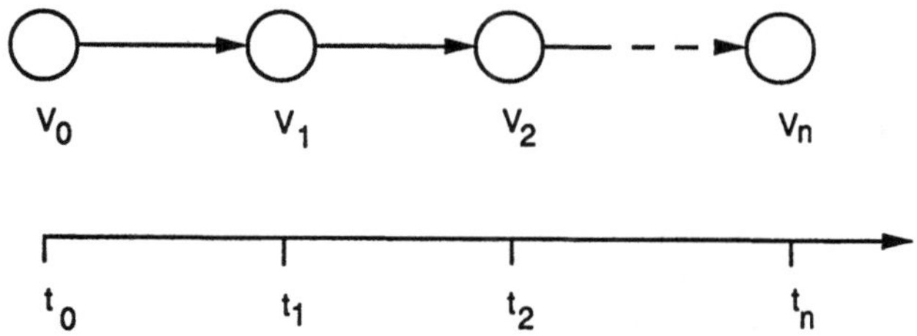

Fig. 3.11. Sequential version generation

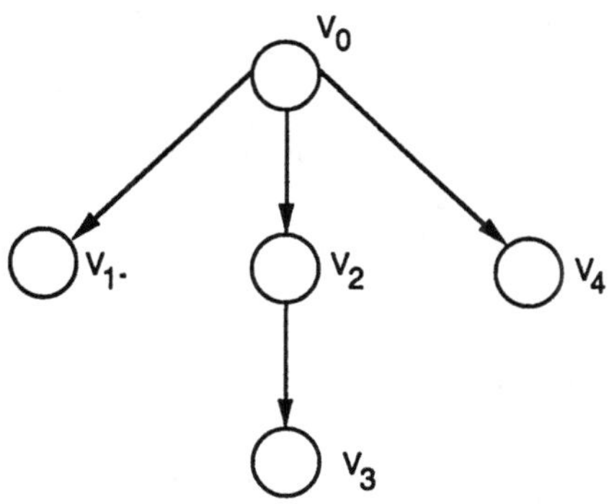

Fig. 3.12. Treelike version generation

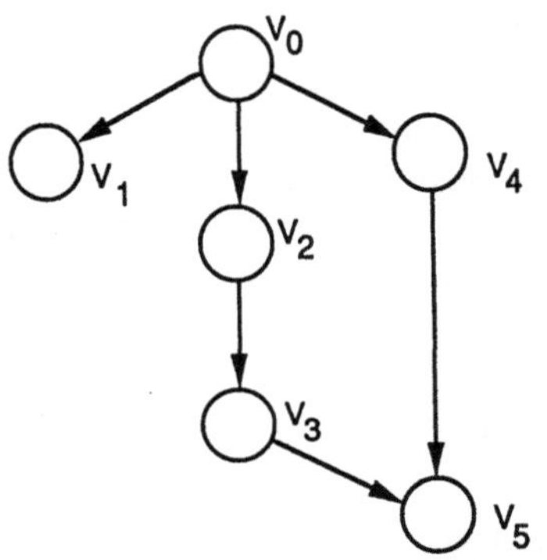

Fig. 3.13. As directed acyclic graph

To distinguish between variants and revisions a tree-like modeling scheme is useful. Figure 3.12 gives an example of a tree-like scheme that models both variants and revisions.

Again we assume that v_0 is the generic version of the object to be designed. Versions v_1, v_2, and v_4 are variants, i.e., they are alternative redesigns of version v_0. Even if we assume that version v_1 is a dead end, in the course of a design process it has to be interpreted as a variant. Version v_3 is a revision of version v_2. Generally speaking, we can interpret a tree-like modeling scheme like this: A certain path in the tree reflects revisions of an object, while different paths reflect alternative solutions. If the numbering of versions implies their creation times, it is possible to manage history information by a tree-like scheme, as well.

Tree-like schemes are useful to decide between variants and revisions. But how do we manage merging versions by a tree-like scheme? In the course of a design process, often the results of different versions are merged to create a new version. This leads us to a directed acyclic graph as a modeling scheme, as in Fig. 3.13.

The graph given above may be interpreted as follows. Version v_0 is the generic version. Versions v_1, v_2, and v_4 are variants, while version v_3 is a revision of version v_2. Version v_5 – the final result – is a merger of versions v_3 and v_4. Hence it is a revision of version v_3, as well as a revision of version v_4. As with a tree as the modeling scheme, history information is managed by a directed acyclic graph, provided the numbering of versions implies their creation times.

To sum up, we have to bear in mind three modeling schemes:

- a linear modeling scheme,
- a tree-like modeling scheme, and
- a directed acyclic graph.

A directed acyclic graph is the most powerful scheme. Using such a scheme, we may not only distinguish between revisions and variants, but also support merging of versions as well as history information. This does not mean that using such a graph is always suitable. The modeling scheme to be used depends on the version interrelations that a designer wants to manage.

A very rough rule is that the modeling scheme to use depends on the particular design phase. It seems useful to have a tree-like modeling scheme during the conceptual design, a directed acyclic graph during the logical design, and a linear modeling scheme for the physical design.

3.3.3 Configuration Modeling Based on Version Management

As discussed in the preceding sections, designs evolve over time. New versions of objects are generated and linked to the corresponding modeling scheme. When looking subsequently at complete systems, they do not consist of just one object, but usually contain many design objects. Setting up a complete system requires picking and incorporating one version per object. This process is called the configuration of a system. The result of a configuration is a system fulfilling a certain functionality and running in a certain environment. Of course, only released versions should be used for a configuration. Figure 3.14 gives an idea of the several possibilities to configure a system.

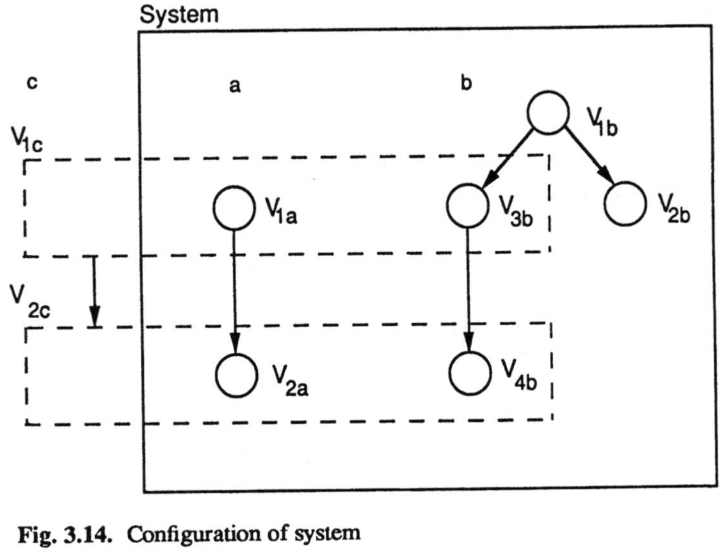

Fig. 3.14. Configuration of system

In this example, a system consisting of two objects is illustrated: object a has two instances – versions v_{1a} and v_{2a} – while object b has four instances – versions v_{1b}, v_{2b}, v_{3b}, and v_{4b}. Since versions v_{1b} and v_{2b} have not been released, they cannot be used to configure our system. In our example, two system configurations are shown:

- a configuration v_{1c} that is set up by versions v_{1a} and v_{3b}, and
- a configuration v_{2c} that is set up by versions v_{2a} and v_{4b}.

Since all configurations of a given system refer to the same set of generic objects, we may also observe interrelationships between versions. These may again be modeled as version interrelations. To manage configurations we have to create a configuration object. Components of a certain configuration are the versions of the design objects picked up and incorporated during the configuration process.

3.3.4 Support of Design Control

Only an engineering database system that offers integrated version management has the capability to support adequate control of the design process. Version management supports:

- structuring of the design process itself,
- management of design interrelations and history information,
- configuration modeling, and
- protection of the data against unauthorized deletion.

We shall briefly examine how some of these mechanisms may affect design control. To start with the last, change or deletion of released versions should be prohibited, otherwise unauthorized changes or deletions may propagate into other released objects. On the other hand, this forces the designer to be very careful with the release of versions. Released versions set up a working base for many other designers. Hence, a version should be released only after passing a battery of verification checks for consistency. On the condition that these checks are always applied, released versions set up a design checkpoint, thus allowing at any time the restart of the design process based on released versions. Additionally it gives a project supervisor the opportunity to check which design parts are already functioning well and which parts are not finished yet.

Any designer with some experience at hand knows about the problems concerning configuration management, e.g., to give an answer to a question like what configuration is installed in what environment. The integration of version management into the database system also adds the capability to handle configurations as indicated in Sect. 3.3.3, i.e., the database system is able to model configuration interrelations and to satisfy retrieval inquiries by the standard query language.

Predefinition of modeling schemes forces designers into a structured design. In this way it becomes much easier to control and to trace the design process.

3.3.5 Version and Configuration Management Based on DAMOKLES

We now return to our example of a VLSI cell. In Sect. 3.2 we outlined a database schema design including the object and relationship definition based on the DAMOKLES model. Problems of version management were not yet included there, thus we have to improve our schema design and our definitions to cover these problems as well.

The data model of DAMOKLES offers all the capabilities of version management that we introduced in the previous subsections: linear version modeling, tree-like version modeling and modeling of versions by an acyclic directed graph. Hence, we will restrict our discussion to this data model.

Let us assume that an engineer wishes to model the cell versions by an acyclic directed graph. Suppose further that whenever a new cell version is set up, a special documentation called 'simulation results' has to be generated. Thus, we have to expand our schema design as illustrated in Fig. 3.15.

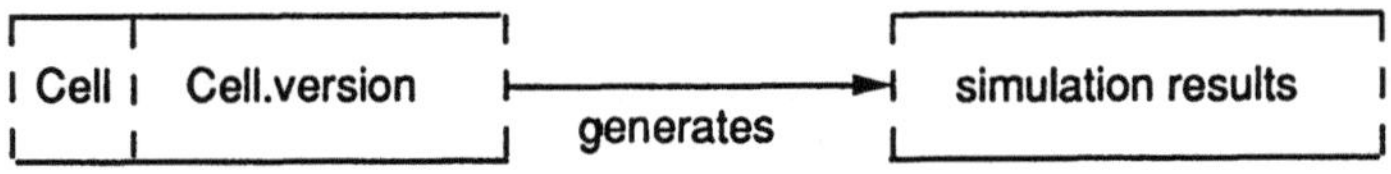

Fig. 3.15. Type declarations for versions

```
OBJECT TYPE cell
  ATTRIBUTES
       .
       .
       .
  STRUCTURE IS
       .
       .
       .
   VERSIONS ACYCLIC
   (
   ATTRIBUTES
       updated : date;
       vname : version name;
    AT LEAST ONCE (generates)
   )
END cell


OBJECT TYPE simulation results
   ATTRIBUTES
       gendate : date;
       results : long;
END simulation results

RELATONSHIP TYPE generates
    RELATES
    ( cell.version,
     simulation_results )
END generates
```

Fig. 3.16. Schema design with versions

The corresponding definitions for the schema mentioned above would roughly look like those of Fig. 3.16. They demonstrate that all we had to do was to expand our cell definition by the special version attributes. We didn't have to create a completely new definition. When looking at the definition of our relationship 'generates', bear in mind that only cell versions are involved, not the cell itself!

Let us now assume that cells are not the only parts of an engineering design, but are also embedded in a system including simulation programs. Of course, there exist a number of simulation program versions that may be organized into a linear schema. Thus, we can again improve our database schema design, resulting in Figs. 3.17 (graphical representation) and 3.18 (type declarations).

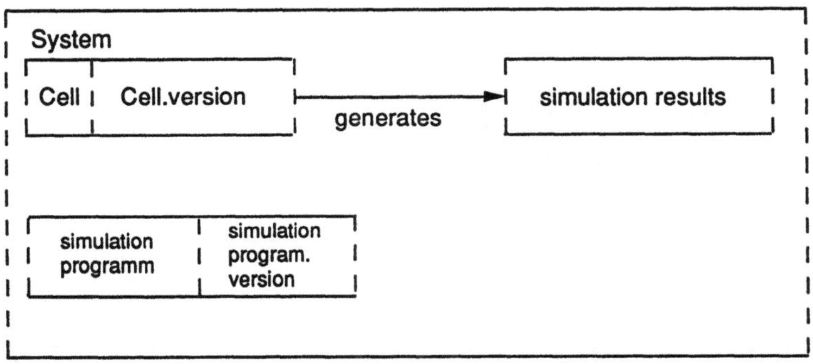

Fig. 3.17. Extension of schema in Fig. 3.15

```
OBJECT TYPE system
  ATTRIBUTES
     sysname : name ;
        :
  STRUCTURE IS
     cell , cell. VERSION,
     simulation_results,
     simulation_program,
     simulation_program.VERSION
 END system

OBJECT TYPE simulation_results
   ATTRIBUTES
      author : name;
         :
   VERSION LINEAR
   (
        ATTRIBUTES
            updated : date;
   )
END simulation_results
```

Fig. 3.18. Type declarations for Fig. 3.17

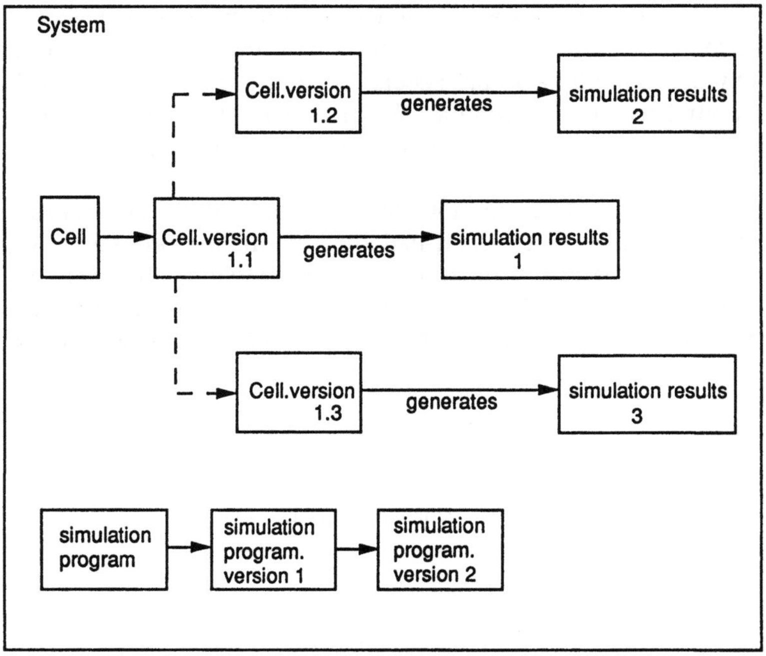

Fig. 3.19. Objects and versions on instance level

Even if our example represents a very small system, it is sufficient to illustrate how to define versions of objects using DAMOKLES. We will now look at the level of design object instances in order to study how the system evolves over time. This might bring us to a situation as in Fig. 3.19.

The figure illustrates samples of objects, object versions, and their relationships in a system under development, and conforms to the schema of Fig. 3.18. Of course, our example contains only a few such objects and object versions. In a real engineering design, a large number of objects and versions are produced. Now, a configuration of a system only contains a portion of all the samples. A particular configuration of a system generally consists of just one version of a particular object. In our example, a configuration may consist of 'simulation_program.version 1', 'simulation_results 2', and 'cell. version 1.2'. Another configuration may contain 'simulation_program.version 2', 'simulation_results 1' and 'cell.version 1.1'.

As mentioned earlier, we may utilize the mechanisms for managing versions also to manage configurations. Different configurations of a particular system are not fundamentally different objects, rather they are versions of the same configuration object. Figure 3.20 illustrates this for the instance level, Fig. 3.21 outlines the type definition for our configuration object.

To summarize, all configurations of a particular system refer to the same objects. To manage configurations, we define a configuration object for each system. Components of a particular configuration are the versions of the design objects picked and incorporated during the configuration process. The configuration interrelations themselves can be modeled like version interrelations.

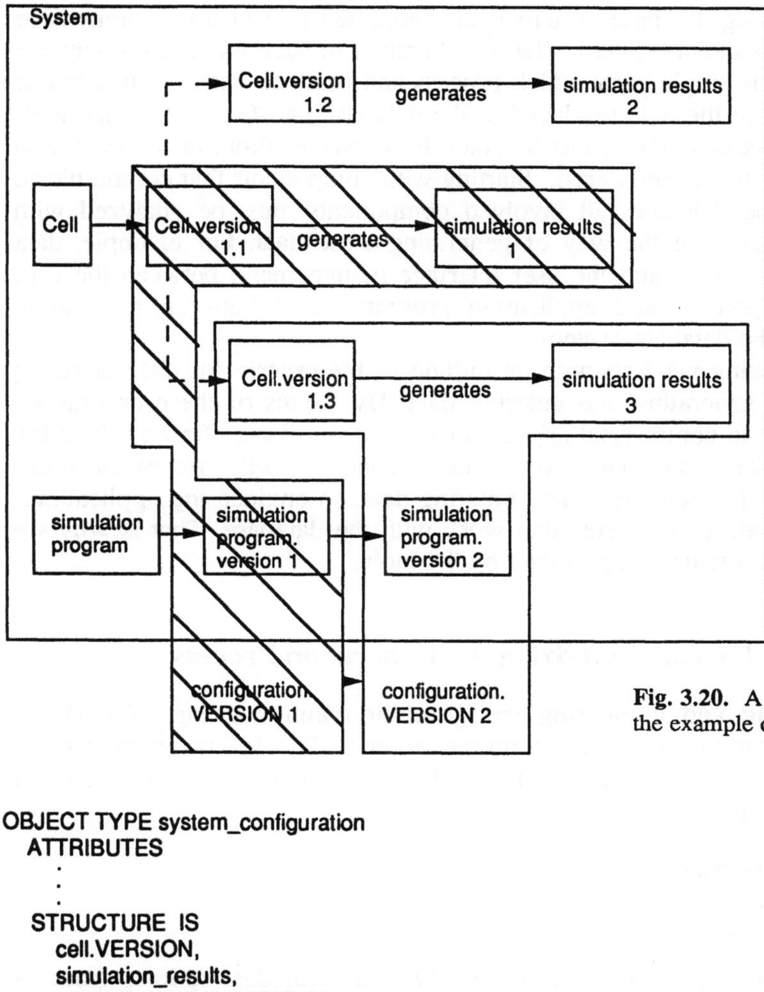

Fig. 3.20. A configuration from the example of Fig. 3.18

```
OBJECT TYPE system_configuration
  ATTRIBUTES
     .
     .
     .
  STRUCTURE IS
     cell.VERSION,
     simulation_results,
     simulation_results.VERSION
  VERSIONS LINEAR
  (
   ATTRIBUTES
     .
     .
     .
  )
END system_configuration
```

Fig. 3.21. Type declaration for configurations

3.4 Generating and Entering Data

In this section, we will analyze in detail the procedures for generating data and entering them in the database. Much of this section deals with the way of generating and entering data by different applications.

An engineering database system is an integrated part of a comprehensive engineering application system. The data handled in such an environment are dependent on the single parts which process and produce data. Although our main interest is in the data produced at the database interface, all components involved in processing data must be taken into consideration; thus, the system architecture has to be considered. Starting with components that communicate directly with the database, all involved components must be analyzed with their kind of data and the way of generating these data. For example, data arise between the user and the user interface management, between the user interface management and application programs, and between application programs and the database system.

In the following we start with an outline of the criteria for characterizing the process of generating and entering data. By means of these criteria we treat the subject in commercial applications. Then we discuss how engineering applications differ from them. As a consequence, we will discern the interfaces important for generating and entering data. In engineering applications, tools are the main components that work with the database. That is why we analyze in detail the handling of data by the tools.

3.4.1 Criteria for Characterizing the Generation Process

When analyzing and comparing the generation and entering of data for different applications, we need common criteria. The following points are used for later discussion of data handling by applications, both in commercial and engineering environments:

- process of generating data,
- kind of data, and
- volume of data.

The *process* of generating data describes the path data follows from its origin to the database interface. Points of interest are the nature of the application (dialogue or batch process, complexity of data processing) and the influence of a user (degree of interaction, use of input devices).

With the *kind* of data we denote whether the application deals with simple attribute values or with structured objects. An example for the first kind are flat tuples in a relational database. Data of the second kind can be objects which are composed of other subobjects or of relationships. The kind of data can be changed by the application (a graphic structure used by an application may be transformed into a long field for entering it into the database).

To describe the *volume* being entered into a database, both size and number of units must be considered. Whereas the size of units specifies whether attributes, tuples, long fields or complex objects are being entered, the number of units is a measure for the number of distinctive entities produced in a design step.

By means of these criteria, generating and entering of data in typical applications can be analyzed.

3.4.2 Commercial Applications

Generally, applications using a database in a commercial environment do not differ very much. The components in a commercial system and their use are shown in Fig. 3.22.

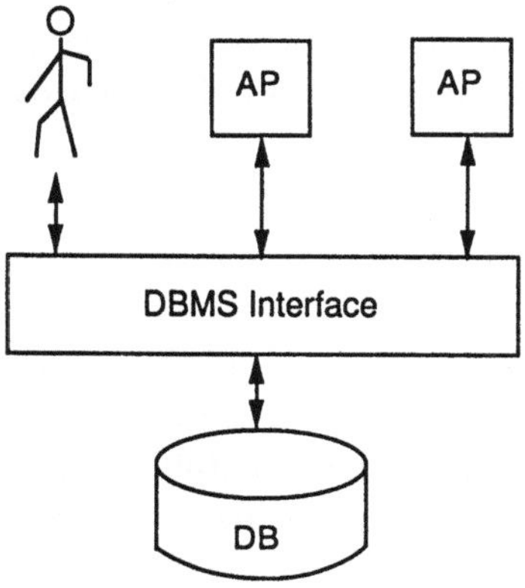

Fig. 3.22. Commercial applications

The following kinds of applications usually run in such systems:

- direct database operations (e.g., in an interactive query language),
- simple interactive applications (e.g., forms languages),
- batch applications, and
- conversion programs.

In *direct operations* the *process* of generating data is very simple. Users refer to simple data structures and data types, i.e., entity types with attribute types and relationship types, defined by the schema, and enter directly the data to be stored in the database. Consequently, the *kind* of data is usually simple attributes. The *quantity* of data entered at one time (DML session) is usually very small, both in unit size and number of units.

Applications with *simple* interactions, e.g., flight booking or banking transactions, use prescibed actions where the data are obtained by filling out a screen mask. The *process* of such programs is retrieving data from the database, making some calculations, and updating the database. The *kind* and *quantity* of data in interactive commercial applications are comparable to direct applications.

Batch applications include monthly salary, statistics, or similar applications. In such a batch *process*, many data can be retrieved from the database, calculations done, reports produced, and also many updates or store operations done. As before we have only simple *kinds* and small unit sizes of data. The *volume* entered into the database is large though, because of the large number of units.

In every database environment, some data conversion programs exist that transform information already existing in machine-readable form into the structure needed for the defined database schema. Such conversion *processes* usually run in batch mode. The *kind* of data generated in these programs are simple, because data types are basic and structures are predefined. The *volume of data* is generally very large (consider, e.g., the number of units in an employee database).

Let us now summarize the facts for generating and entering data in commercial applications. In most cases, there is a simple process to generate the output data from some input. Sometimes data is generated interactively by the user. There are also simple kinds, and very often, small volumes of data to be handled. Only in batch processes will the volume of data grow. In all cases the size of units is small.

3.4.3 Engineering Applications

In engineering applications, data are usually generated and processed by software tools. Database systems are used by applications programs to manage their data. That means that the user accesses the database mainly via the tools. The user's access to tools takes place via a uniform user interface. We assume that he does not work directly with the database. The database will be transparent for users via user interface management and application programs. This yields, compared with commercial applications, a completely different system architecture. Figure 3.23 shows the rough architecture of an integrated engineering design system. We can discern three main parts:

- user interface management and device-independent graphic system,
- database management system, and
- application modules.

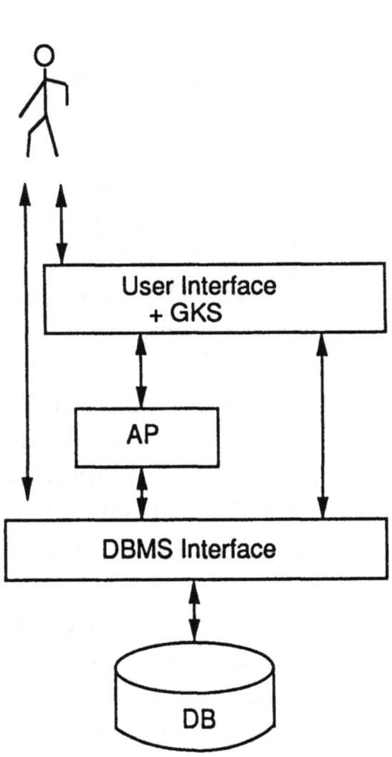

Fig. 3.23. Engineering applications

We briefly describe these parts and afterwards enter into a discussion of the different data flows between the components. The *user interface management* realizes the dialogue between user and application modules. The *device independent graphic system* (graphical kernel system, GKS) manages the representation of graphical output produced by the application or directly extracted from the database. It is also responsible for graphical input in cooperation with the user interface management. The *database management system* is the central part in the design environment. The *application modules* comprise all software tools required for the design. This architecture should not be seen as a rigid structure. When we show user interface management and the graphical kernel system as a single component it does not mean that both cannot be divided into two parts. In this chapter, only the existence of such components and their data communication is important. Consequently, not all the data flows present in every design environment will be discussed below, and not all of those discussed may be found in a given system. Their presence depends on the configuration and the systems integration. Interfaces exist between:

- user and database management system (DBMS),
- user and user interface management/GKS (UIM/GKS),
- UIM/GKS and application modules,
- UIM/GKS and DBMS, and
- application modules and DBMS.

The interface between *user and DBMS* serves mainly for direct querying to the DBMS by the designer. The user's most frequent interactions are performed via the *interface to UIM/GKS*. The interface between *UIM/GKS and application modules* is necessary to control the user's applications and graphic representations of application results. The last two interfaces, *UIM/GKS and DBMS*, and *application modules and DBMS*, can be treated alike, if we take the UIM/GKS as an application from the database point of view.

To study the latter in more detail, the communication of the tools with the database must be analyzed. An engineering system consists of three classes of tools:

- synthesis tools,
- analysis tools, and
- tools for design information management.

In the first class we find *synthesis* aids for design capture, which transform the design description into machine-processable form. The simplest methods are to digitize a pencil and paper specification, or to use a language-based text editor for creating some program code. In VLSI design we find integrated circuit (IC) geometry editors for specifying the processing masks for circuit fabrication, or schematic editors for creating a description of a hardware system at the logic gate level.

The *process* is marked by high human interaction, often by using graphic input devices. The application programs transform human input (graphic input) to the internal representation, which can be – depending on the data base schema – complex objects or long fields, which determine the *kind* of data for the database.

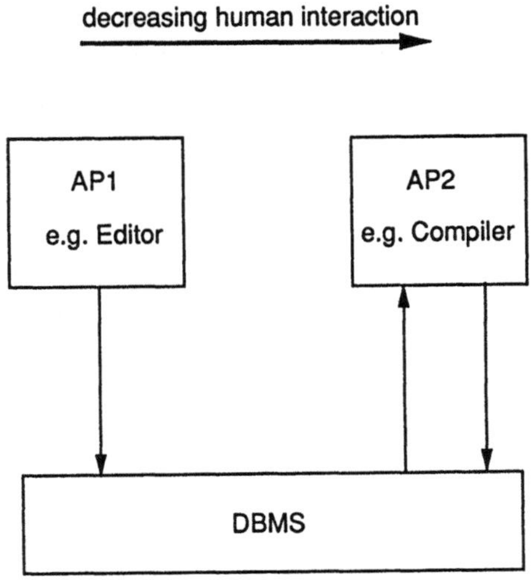

Fig. 3.24. Human interaction

Geometric description of designs produce a large data *volume*. Both the units to be entered into the database (complex objects or long fields) and the number of units (e.g., many transistors) are large.

Among synthesis aids are also mapping tools that map the high level description of an object into its physical implementation. Examples for this type are programming language compilers or module generators in software engineering, and PLA generators in VLSI design. Placement and routing tools, which place subobjects in a two- or three-dimensional space and interconnect them, are also part of the synthesis class. All these tools generate data like machine-language instructions, fabrication-mask geometries, wrap lists or printed circuit board (PCB) artwork, and ship or building blueprints. A designer interacts much less in these kinds of *processes*. We detect decreasing human interaction (see Fig. 3.24).

The *kind* of data is comparable with the last class. However, note an increase of the *volume* (e.g., stick diagrams are being expanded to physical layouts).

Another important set of tools deals with the *analysis* of designs. A simulator models the behavior of a system by using an abstract description of it. In most cases the model can be formulated mathematically. With topological analysis tools the correctness of the layout and interconnections of subsystems can be checked. An example is a geometric design rule checker in VLSI design. A third class of analysis aids contains timing analyzers for finding critical timing paths in VLSI designs, or for profiling the run times of a program in software engineering.

When examining the data handled by this class of applications it is useful to distinguish between dynamic and static analysis. Tools for static analysis like topological analyzers or timing analyzers take a particular design representation as input and produce a report describing the errors encountered, a list of critical paths, etc., as output. The analysis is independent of the designer's input; it uses representational data produced in another, earlier step of the design. Dynamic analysis tools, which comprise all kinds of simulators, produce outputs which are critically dependent on an input of test data. Therefore, the system must manage large quantities of data that are separate from representational data, such as input of test data cases and the corresponding output data. The resulting output must also be stored in the database for later use, e.g., in the test phase, or for redesign.

To conclude, whereas static analyzers exclusively *process* data produced by an earlier design step, dynamic analyzers need some additional data like test data or stimuli. These data can be obtained by human interaction or from a database. Applications without human interaction can run in batch. Analysis tools produce a large data *volume* which must be stored in the database for later use.

The third class consists of tools for *design information management*. An engineering design system must provide facilities to query for information about general design management data like status of design parts, versions, releases, or designers involved in a design. Furthermore, the design management must guarantee the consistency of designs by starting complex consistency check applications.

The *process* of design information management is highly interactive, and typically receives input data of simple *kind* and *volume* that resemble those of commercial applications.

3.4.4 Example

In Sect. 3.2 we designed a database schema for our VLSI example using the DAMOKLES data model. Now let us proceed with generating concrete objects and inserting them in a DAMOKLES database. That means we have to create the cell prototypes CELL1, CELL2, and CELL3, the appropriate ports and – for CELL3 – the subobject instances. Furthermore, the connection between the different ports must be inserted.

For the purpose of illustration we identify the ports by name (as indicated in Fig. 2.21), using the following conventions: Input ports are named *A*, *B*, *C*, *D* and output ports are named *X*, *Y*, *Z*.

DAMOKLES works similar to the description of Fig. 2.32, in that its operators reference buffers (work areas) containing data for insertion (e.g., attribute values). We use in the following examples a PASCAL-like syntax for handling buffers. For example, the expression *cell_buffer.name* stands for the component *name* of a structured record *cell_buffer*.

The first step in entering the objects of the example into the database is the insertion of cell prototype objects. We use the DAMOKLES function **db_io** to insert CELL1, CELL2, and CELL3 into database vlsi according to Fig. 3.25. The database system returns the database key of the object, assigned to the variables *k_CELL1*, *k_CELL2*, and *k_CELL3*, respectively. This key can be used for later references to the object (e.g., when inserting subobjects).

The next step is to add ports to cell prototypes. This is performed by inserting ports as subobjects of cells with the function **db_io** again (Fig. 3.26).

```
/* insert cell prototype CELL1*/
cell_buffer.name := `CELL1`;
cell_buffer.released := TRUE;
db_io (IN vlsi,cell,cell_buffer,OUT k_CELL1.err_code);

/* insert cell prototype CELL2 */
cell_buffer.name := `CELL2`:
cell_buffer.released := TRUE;
db_io (IN vlsi,cell,cell.buffer.OUT k_CELL2,err_code);

/* insert cell prototype CELL3 */
cell_buffer.name := `CELL3`;
cell_buffer.released := TRUE;
db_io (IN vlsi,cell,cell_buffer,OUT k_CELL3,err_code);
```

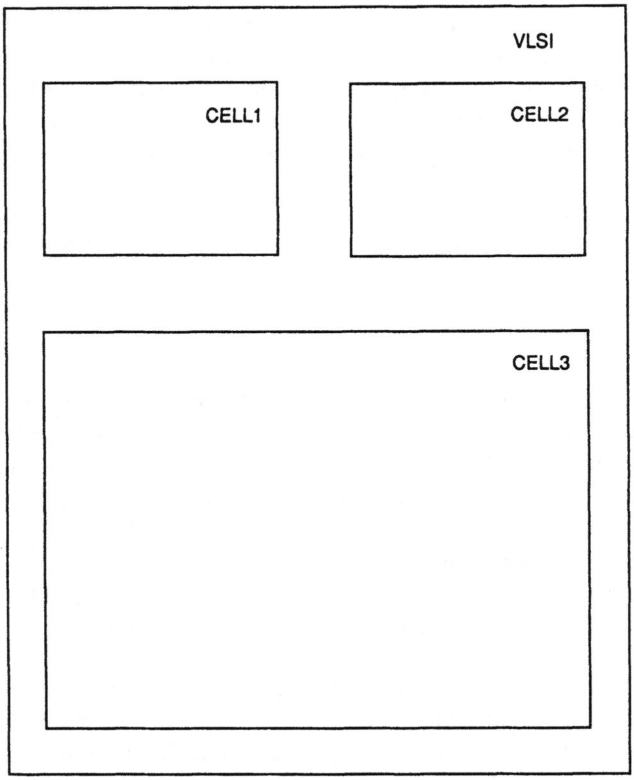

Fig. 3.25. Insertion of cell prototypes

The port buffer *port_buffer* is filled with the name and the direction of the port. Then the *port* object is inserted into the database *vlsi* as a subobject of a *cell* object. For port *A* of *CELL1* the *k_CELL1* key is specified to identify the port as a subobject of *CELL1*. The return key must be saved for making connections later on.

Now the prototypes CELL1 and CELL2 are fully specified. The exact internal implementation is not necessary for our example. However, for CELL3 we have to generate the concrete realization, i.e., insert the instances

```
/*insert ports of cell prototypes CELL1 */
port_buffer.name := `A`;
port_buffer.i_o. := `INPUT`;
db_io (IN vlsi,port,cell,k_CELL1,port_buffer,OUT k_CELL1_A,err_code);
port_buffer.name := `B`;
port_buffer.i_o. := `INPUT`;
db_io (IN vlsi,port,cell,k_CELL1,port_buffer,OUT k_CELL1_B,err_code);
port_buffer.name := `X`;
port_buffer.i_o. := `OUTPUT`
db_io (IN vlsi,port,cell,k_CELL1,port_buffer,OUT k_CELL1_X,err_code);
port_buffer.name :=`Y`;
port_buffer.i_o. :=`OUTPUT`;
db_io (IN vlsi,port,cell,k_CELL1,port_buffer,OUT k_CELL_Y,err_code);

/*insert ports of cell prototype CELL2 */
port_buffer.name := `A`;
port_buffer.i_o. := `INPUT`;
db_io (IN vlsi,port,cell,k_CELL2,port_buffer,OUT k_CELL2_A,err_code);
    .
    .
    .

/* insert ports of cell prototype CELL3 */
port_buffer.name := `A`;
port_buffer.i_o. := `INPUT`;
db_io (IN vlsi,port,cell,k_CELL3,port_buffer,OUT k_CELL3_A,err_code);
    .
    .
    .
```

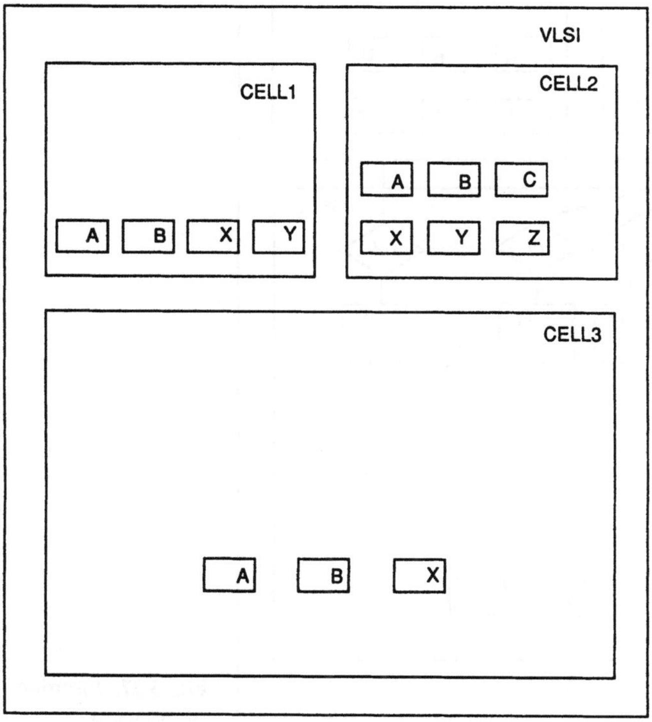

Fig. 3.26. Insertion of ports

```
/* insert instance I1 */
inst_buffer.name := `I1`;
inst_buffer.creation_date := .....;
db_io (IN vlsi,instance,cell,k_cell3,inst_buffer,OUT k_I1,err_code);

/* insert reference to prototype */
roles[1].key := k_I1;
roles[2].key := k_CELL1;
db_ir (IN vlsi,ref,cell,k_CELL3,roles,OUT k_...,err_code);
    .
    .
    .

/* insert instance I5 */
inst_buffer.name := `I5`;
inst_buffer.creation_date := ....;
db_io (IN vlsi,instance,cell,k_cell3,inst_buffer,OUT k_I5,err_code);

/* insert reference to prototype */
roles[1].key := k_I5;
roles[2].key := k_CELL2;
db_ir (IN vlsi,ref,cell,k_CELL3,roles,OUT k_....,err_code);
```

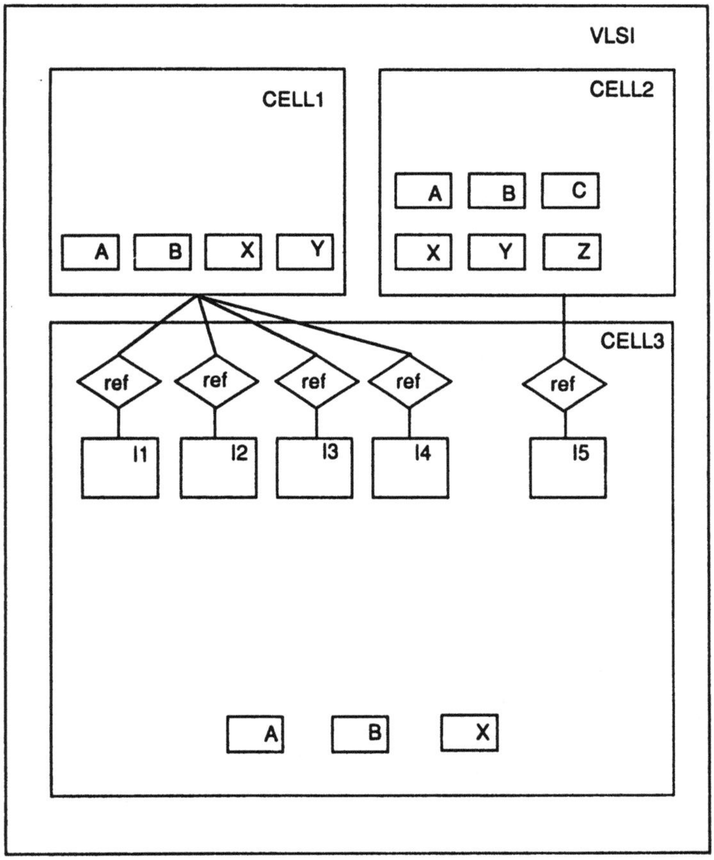

Fig. 3.27. Insertion of instances

```
/* insert port instances for I1 */
/* copy cell prototype port into instance */
db_cpo (IN vlsi,port,k_cell1_A,instance,k_I1,DB_DESC,OUT k_I1_A,err_code);
db_cpo (IN vlsi,port,k_cell1_B,instance,k_I1,DB_DESC,OUT k_I1_B,err_code);
db_cpo (IN vlsi,port,k_cell1_X,instance,k_I1,DB_DESC,OUT k_I1_X,err_code);
db_cpo (IN vlsi,port,k_cell1_Y,instance,k_I1,DB_DESC,OUT k_I1_Y,err_code);

/* insert port instances for I2 */
db_cpo (IN vlsi,port,k_cell1_A,instance,k_I2,DB_DESC,OUT k_I2_A,err_code);
    :
/* insert port instances for I5 */
db_cpo (IN vlsi,port,k_cell2_A,instance,k_I5,DB_DESC,OUT k_I5_A,err_code);
    :
```

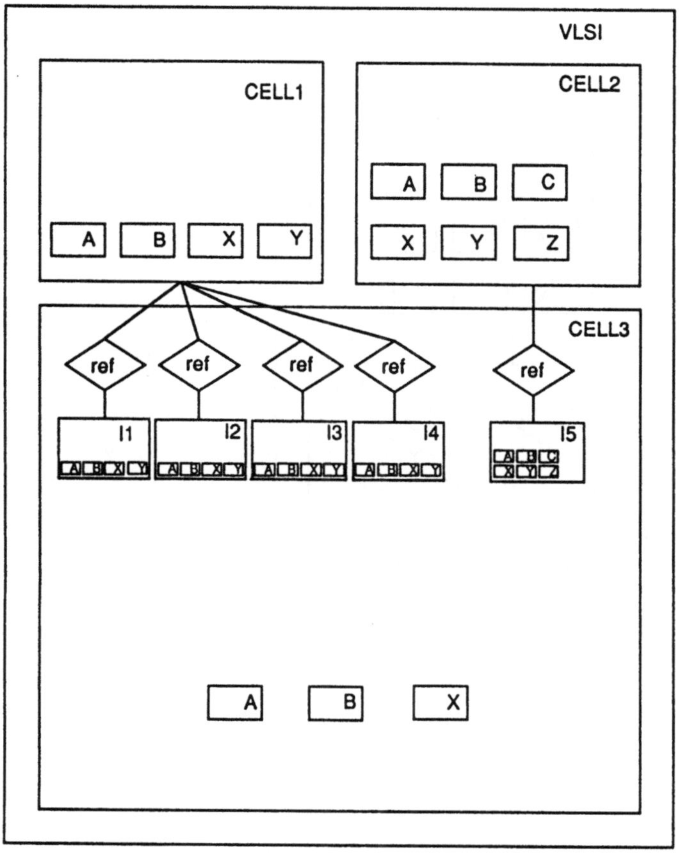

Fig. 3.28. Insertion of port instances

building the composition and the connections between ports and port instances. The function **db_io** is used again to insert objects of type *instance* into cells. We have only said, by this action, that a cell is composed of some instances, but we have never specified from which prototype the instances are built. We can do this by inserting a relationship between instances and prototypes. The function **db_ir** inserts a relationship of type *ref* as subobject of a

```
/* insert connections of ports */
/* connect port A of CELL3 with port instance A of I1 */
roles[1].key := k_CELL3_A;
roles[2].key := k_I1_A;
db_ir (IN vlsi,connect,cell,k_CELL3,roles,OUT k_...,err_code);

/* connect port instance Y of I4 with port X of CELL3 */
roles[1].key := k_I4_Y;
roles[2].key := k_CELL3_X;
db_ir (IN vlsi,connect,cell,k_CELL3,roles,OUT k_...,err_code);

/* connect port instance Z of I5 with port instance B of I4 */
roles[1].key := k_I5_Z;
roles[2].key := k_I4_B;
db_ir (IN vlsi,connect,cell,k_CELL3,roles,OUT k_...,err_code);
   :
```

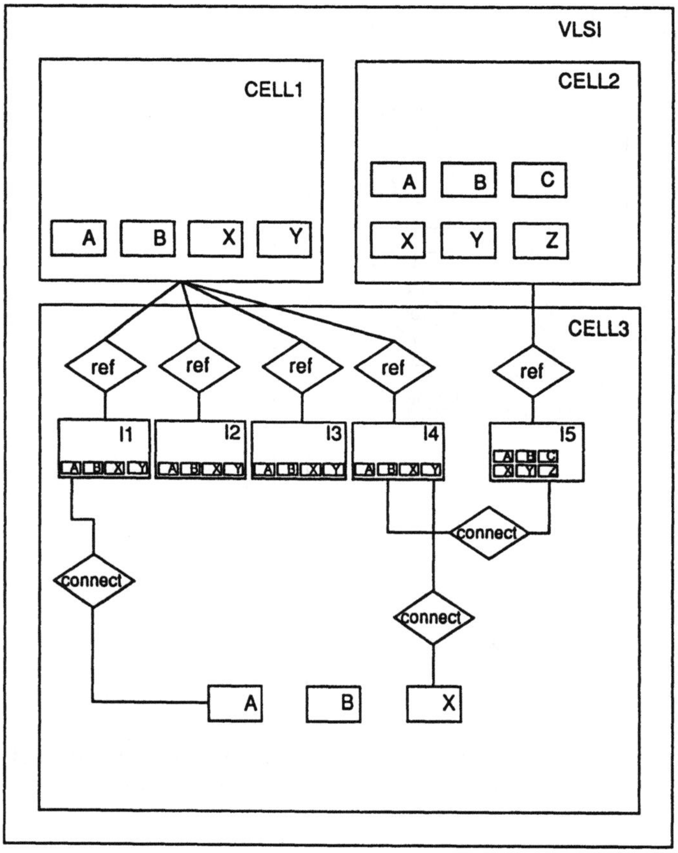

Fig. 3.29. Insertion of connections

cell object. Thereby the database buffer *roles* contains the database keys of the objects standing in relation (e.g., *k_I1,k_CELL1*). The generation of CELL3 instances is shown in Fig. 3.27.

Normally, instances inherit the ports of their cell prototype. However DAMOKLES does not support this constraint. Therefore, the application is responsible for the correct insertion of port instances. There does not exist any mechanism in the database system to guarantee this integrity. One of the shortcomings of DAMOKLES, the lack of an inheritance mechanism, also points to the importance of inheritance via the ISA-relationship. However, the correct insertion of port instances will at least be supported by the existence of a copy operator. Figure 3.28 gives an example for inserting port instances explicitly. We use the function **db_cpo** to put into the database a copy of the prototype's ports as subobjects of the corresponding instance.

As a last step, the ports must be connected. We can discern, generally three kinds of connections:

- prototype port connected with prototype port,
- instance port connected with instance port, and
- prototype port connected with instance port.

A relationship *connect* is established between connected ports and inserted as a subobject of a cell prototype. The roles involved are the prototype ports and instance ports (Fig. 3.29). Besides, instance ports must be qualified with the cell name to guarantee a unique identification.

Some comments about the use of these functions are in order. The functions are not intended to be used by a normal designer. Instead, they serve as an interface for design tools like layout editors. When, e.g., the user of such an editor designs a cell, the application program must perform the appropriate database functions. The user does not perceive anything about the calling of these functions, so that their unwieldiness is of no consequence.

3.5 Archiving of Database Objects

The main objective of current hardware and software engineering systems is to support the design process during all phases of the system life cycle by integrating existing and new tools in a common environment. Integration of different tools is achieved by adapting them to a common database interface and a uniform user interface.

A database system as part of a system engineering environment should not only give support for version and configuration control, but should also include a powerful archiving subsystem which supports the large amount of data generated during a long period of development. Archiving is controlled by the user in units of the data model, e.g., complex objects and configurations. Candidates for archive are data less frequently used, or those which must be kept off-line for security reasons. Versions and configurations usually belong to the former category of data because they originate in various phases of the design life cycle, and have to be kept for a long period of time, although typically they are seldom accessed.

Whereas the integration of concepts for complex objects and for version management into one data model have been extensively studied (see Dittrich et.al.86, Dadam et.al.84, Lorie et.al.83, and Katz85, among others), there seems to be as yet no concept or realization of fully integrating an archive component into a database system in a way which adds a capability at the database user interface level. One of the first ideas for using optical disks as a storage medium for large file versions was proposed by Katz in [Katz et.al.84] who archives versions automatically without user input. Katz treats the archive only as storage extension on the physical system layer. In contrast, a database archive should be based on concepts where the user explicitly directs the archiving of objects.

This section describes how an archiving component may deal with versions and configurations of objects. Starting from requirements like integration aspects for an engineering archiving system, some special archiving operations are described by pointing out the functionality for archiving versions and configurations. As a point of illustration we shall use PRODAT, a database system that serves an environment for developing real-time software.

3.5.1 General Requirements for an Archiving Mechanism in a Database System

The original idea of archiving in the context of data processing was developed for storing ordered deposits on an inexpensive, high-capacity storage medium. Longer access times and even unavailability – if the medium was off-line – were accepted (grudgingly). More features are desirable, however, in the archiving component of a modern database management system.

PRODAT focuses primarily on the integration of archiving operations into the data model. That is, archiving is done in terms of simple and complex objects as opposed to a mere dump of the database files. Information about the archiving status of objects is managed by the database system. It is possible to apply common database operations to archived objects if the corresponding medium is on-line.

Another requirement is that the archive medium should retain its data over periods of decades (instead of about one and a half years as magnetic tapes do). As a consequence, the archiving subsystem uses optical disks as a storage medium which provides durability for thirty years (more recently, even a hundred years are being guaranteed). Optical disks also have high capacity (1 Gigabyte per side) at a low price. Their access times are comparatively high, but tolerable for rare access.

In the next section the archiving units, namely simple objects, complex objects, configurations, and databases are explained in more detail.

3.5.2 Archiving Objects

We start by outlining the PRODAT data model. Complex objects consist of simple objects connected by standard relationships. Simple objects are described by attributes and may contain text, object code, pixel data, etc., which is called the *contents* of the object. The contents are a variable-length string

```
TYPE    logical_object
        STRUCT
        name  CHAR (20)                          /* attribute definition */
        date  DATE                               /* attribute definition */
        CONTENTS                                 /* contents definition */
        SUCCESSORS                               /* subobject definition */
        1 text_object   OR 1..2 picture_object   /* non-exclusive or! */

;
```

Fig. 3.30. Type definition of an object

of bytes which is not interpreted by PRODAT. *Attributes* are similar to those of the relational model, where each attribute has a domain and is unstructured. Take as an example an object that contains a program source code, and the attributes describe the generation date of the program, the author, the programming language, and an internal code number. Figure 3.30 shows part of the corresponding schema definition.

The structure of complex objects is modeled through *relationships* which are defined in the SUCCESSORS part of the schema definition. It is possible either to give an expression that determines the set of admissible successors or to use one of various standard relationships (e.g., has_component or has_alternative). Relationships group simple objects in a hierarchical manner; there is no limit, however, to the number of predecessors of an object.

Completion of complex objects is a necessary condition for closing a phase of computer-aided design process. The structuring of objects might be a long process during which objects are not guaranteed to stay consistent at each state of development. But tools and applications need some predefinitions and restrictions when using objects. It should be possible to fix certain conditions for the structure of complex objects in the schema definition of objects. This is the reason why PRODAT supports definitions about the completeness of complex objects.

A complex object is in the state *complete* if the set of its current subobjects matches the expression in the SUCCESSORS part of the object's definition both in type and cardinality. The completeness status of an object is evaluated by the system each time a subobject is moved or otherwise manipulated.

Take again the schema definition in Fig. 3.30. It determines that for each object of type "logical_object" one subobject of type "text_object", one or two subobjects of type "picture_object", or a combination of both cases must be present for the object to be complete. Incidentally, this defines the has_alternative standard relationship.

Figure 3.31 gives an example of a complex object with subobjects. For certain kinds of combinations of types in the subobject definition part, PRODAT defines standard relationships which are typical structures of objects in system engineering environments. The precise definition of these standard relationships is not necessary in the context of this paper. The example in Fig. 3.31, however, provides some basic ideas as discussed below.

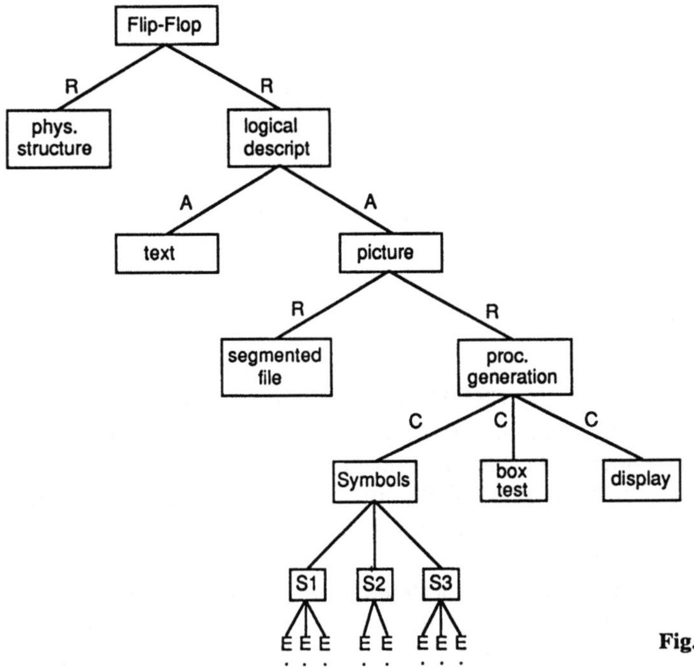

Fig. 3.31. A complex object

The object "Flip-Flop" has two subobjects "phys. structure" and "logical descript.", which are connected to it by the relationship has_representations (R). In this case "Flip-Flop" is complete if the representations both exist and if each of them is of the appropriate type. Object "logical descript." has two subobjects "text" and "picture", which are connected to "logical descript." by a relationship has_alternative (A). The has_alternative relationship defines that the superobject "logical descript." is complete if any one subobject of the types specified exists. Object "proc. generation" has three subobjects, which are connected to it by the has_component relationship (C). The object "proc. generation" is complete if exactly the specified number of subobjects for each defined type exists.

Two special relationships were considered important enough to support them as separate concepts, namely version and configuration dependencies; they are treated in separate sections below.

Let us now consider the archiving of such objects. Simple objects are archived together with their type definitions. Next, all relationships in which the object plays a role are inspected. If the other object involved in a relationship is already archived, the relationship itself will be archived as well. Otherwise the relationship remains in the database. To preserve referential integrity in this case an "envelope" of the archived object is kept in the database, which contains only the object's name and archiving status.

The operation to archive a simple object may serve as an example for the interface of the archiving mechanism in the C language. The C procedure header is given below.

RC archive_simple_object (archive_id, simple_object)
*char *archive_id;*
KEY simple_object;

Archiving complex objects is done in a similar manner. A complex object consists of its root together with all its subobjects plus the relationships, so these elements are archived. Besides, there may be references to an archived subobject from a second complex object. Subobjects referenced in this way, as well as the root itself, are kept redundantly as "envelopes" in secondary storage to maintain referential integrity. Further relationships may have to be established between the complex object to be archived and objects already present in archive.

One reason for including the operation archive_complex_object in the interface is to be in accordance with the object philosophy. Another reason is convenience; the effect of archiving a complex object would otherwise have to be achieved by multiple calls of archive_simple_object.

To archive a database means to transfer into archive all data items of that database left in secondary storage. This operation serves more than mere convenience, but a full explanation is beyond the scope of this paper.

The central idea behind this procedure is to archive *incrementally*. A complex object may have been archived in parts; but it nevertheless constitutes a single entity once it is completely in archive. Only the times at which they were archived will indicate that the parts were archived separately.

3.5.3 Archiving Versions

In PRODAT, versions are objects which are derived from one another. This directed dependency is maintained in a is_derived_from relationship. It is possible not only to have a linear sequence of versions, but also to build version trees by deriving variants from one version which in turn may spread out arbitrarily (version tree).

To create a new version of an object the original object has to be *released*, which means this object cannot be further changed. If it is changed, dependent versions can no longer be derived from it. Only complete objects may be released.

The user can generate versions from both simple and complex objects. If a version is created, a new object is generated automatically and the system connects the original and the new object by the version relationship described above. Both the original and the new object acquire version status; nevertheless, from the user's point of view they are treated like normal objects. The new version is not released and, therefore, can be arbitrarily manipulated.

Figure 3.32 shows a version of a complex object, named "logical descript.1". If the version "logical descript.2" is created as a complex object, the user must specify that the unit for making a version is a complex object (instead of a simple one). Version relationships between the original and new versions are generated automatically. Each new object with version number 2 is now embedded in a different version graph. After this generation process

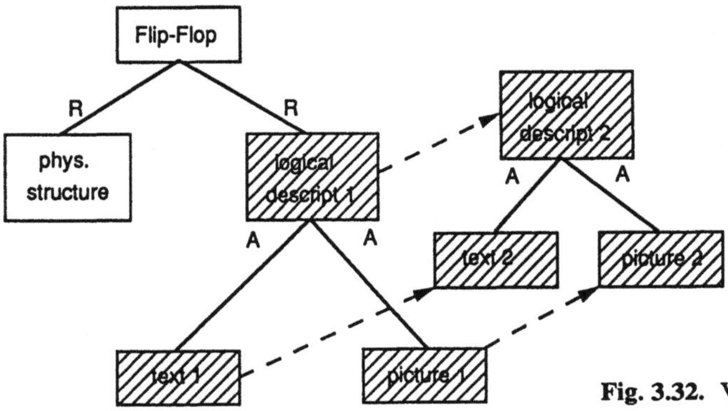

Fig. 3.32. Versions of a complex object

the same standard relationships as in the original object exist between the object and the subobject (in the new version). This is because versions of an object have the same object type, and version generation of a complex object does not change the structure of the origin. Starting from this state, it would be possible to change the structure of the new version object or to manipulate anything else (e.g., the contents).

In summary, every version in PRODAT is also a "normal" object, be it complex or simple. Therefore, no special operations are provided to archive versions. The operations to archive simple or complex objects, however, must support version relationships. Whenever two objects connected by a version relationship are in archive, the relationship must also be present there. So if a version tree is archived in parts, it will be in archive as a whole.

3.5.4 Archiving Configurations

The notion of configuration in PRODAT is slightly more general than the one used in Sect. 3.3. In PRODAT, configurations can be seen as a logical grouping of objects; but it is usually necessary to provide some additional information describing the configuration. For example, a configuration could consist of a customized software package which is taken from a complex object containing all potential parts of a package. Configuration-specific data could then be the name of the customer, the date of delivery, or the price.

Configurations therefore contain an "inventory" of objects together with user-defined attributes for each object. Configurations are described in a special configuration type definition which contains selection rules similar to the completeness rules mentioned above. The selection of objects must happen in accordance with the relationship involved. For instance, *all* children of a chosen object must be picked up (if using the has_component rule) or *exactly one* child must be picked up (if using the has_alternative rule). As a result, the user can influence the configuration process by fixing a certain relationship at type definition time.

Creating a configuration is done by indicating the configuration type desired and the simple objects involved. A conformance test with the configuration type definition is performed automatically. If any rule is violated, the

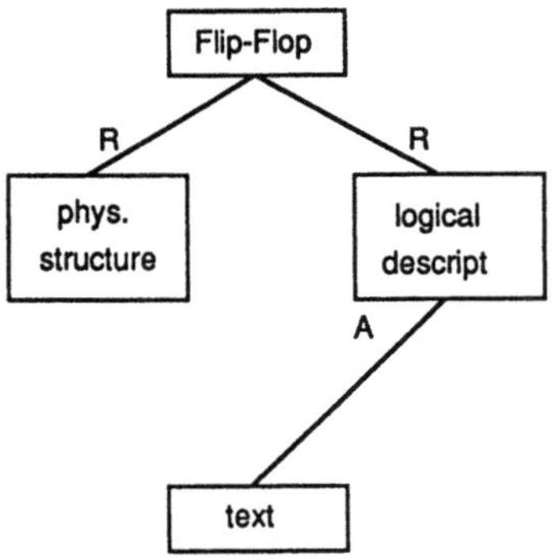

Fig. 3.33. A configuration of a complex object

action will be rejected; the same holds true if any of the objects involved is not released.

Figure 3.33 gives an example of a configuration; it originates from the complex object shown earlier.

The main advantages of a separate configuration concept are as follows: Configuration structure is predefined to a certain degree by type definition, configurations can be treated like "normal" objects, and they can be supplied with extra individual attributes.

Archiving configurations raises more problems than archiving versions. The reason is that configurations reference objects, but do not actually contain them. So it is possible to have an object referenced by many configurations.

Consider an archived configuration. Should the corresponding objects be automatically archived as well? To determine this, the user's view of configurations must be considered. Integrating the functionality of archiving configurations into the database system's set of operations requires that one maintain the logical view and be consistent with other database concepts. Therefore the physical behaviour of the storage system (i.e., keeping references instead of copies) must be hidden as much as possible. As long as only secondary storage is under consideration, there is no problem. Trouble arises with the fact that the archive medium is interchangeable. The following situation gives an example of this problem (Fig. 3.34).

A configuration C1 together with a referenced complex object S is in archive. Another configuration C2 is in secondary storage. C2 also references S.

Assume the current archive medium is full, so that C2 cannot be archived in it. In order to archive C2 the medium has to be changed (assuming there is only one disk drive). However, it is impossible to archive C2 together with S, because S is inaccessible at that moment.

Recognizing the general symmetry in C1 and C2 if there is only one S available, redundancy of S becomes the solution to the problem. In this special case redundancy is not dangerous for consistency, because none of the participating objects will be altered any more.

The following concept was chosen: In archives, configurations are stored together with the objects they reference. These objects are not deleted from the database in secondary storage; the configurations themselves are. From the

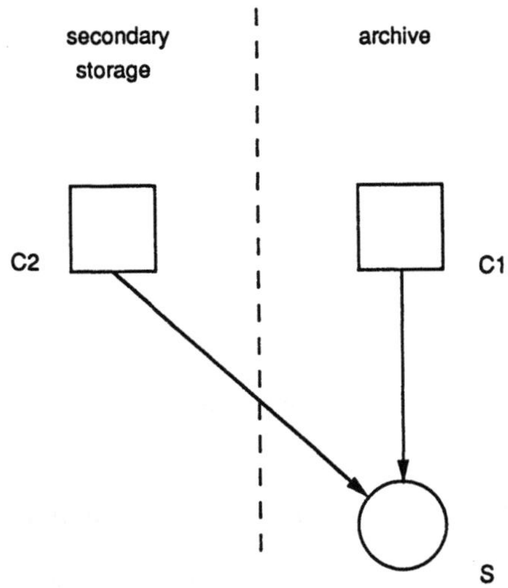

Fig. 3.34. Configuration referencing objects

user's point of view, the referenced objects are not archived and therefore can be used in other configurations. In particular, other configurations can be arbitrarily archived without regard for overlapping configurations.

3.6 Data Interchange

3.6.1 Importance of Data Interchange in Engineering Applications

The interchange of data has been an important subject in engineering environments for many years. In such an environment, data communication takes place between all components. Therefore it is very important to emphasize what exactly we mean by *data interchange*.

Data interchange should not be interpreted as every data flow in engineering applications. The communication between the engineering database and the other components which use it, like application modules, is covered in Sects. 3.4 and 3.7. The interaction by the user is a subject intensively discussed in connection with user interface management and is not a topic of this book.

In this section we will analyze the exchange of data between different components which are not integrated by a common database, and between integrated systems and the external environment or other systems.

Our interest in interchange is not the physical way in which data will be transported. From our point of view it doesn't matter whether the transportation will be done by writing and sending a magnetic tape, or using a local or wide area network. We are only interested in the kind and structure of the data themselves.

Of course, data interchange depends on the level of system integration. Therefore, we will deal at first with the different levels of integration. We will examine the communication between application tools which are not integrated, between integrated systems, and between their environment. Furthermore, we will analyze the data interchange caused by the existence of different phases in engineering design, and different companies or groups involved in design.

After that the respective partners which interchange data will be presented, i.e., the sender and receiver of data.

From these perceptions the kind and the structure of data to be interchanged will be derived. Some known examples from mechanical engineering and VLSI design will be presented.

Then we will see that these existing formats do not fulfill the requirements caused by new comprehensive applications. The shortcomings of these formats and requirements for new formats also in relation to engineering database systems will be presented.

Finally, we will outline a new format for VLSI design, which aims to fulfill these requirements.

3.6.2 Data Interchange in the Different Levels of System Integration

As mentioned above, the data interchange depends on the level of system integration. For further discussion we distinguish three levels of integration:

- collection of application modules, where each component uses its own database,
- one database for all application modules of a single engineering process (e.g., a database for CAD, a database for CAM), and
- integration of all application modules by a common database (e.g., a database for CAP, CAD, and CAM).

These classes are not mutually exclusive; rather, there is a continuous transition from one level to the other. We often have the situation that some applications are already integrated, while others are not yet integrated or can not be integrated for a variety of reasons. The effort for integration of the application may be irritably high, or the part to be integrated is a closed turnkey system.

Let us now discuss data interchange for the three classes described above. In the first class where we have a *collection of application modules*, data interchange must be established between any two applications that rely on each other's data.

Consider a toolbox for VLSI design as an example: The data produced by a schematic entry editor will be used by a logical simulator to check correctness. As the two tools are not integrated by a database, the data must be interchanged by a suitable interchange format like, e.g., SPICE.

In *integrated applications* data interchange is necessary if we want to go outside the integrated system. This may happen if different databases for every engineering phase, like CAD and CAM, exist.

Even if we have a *single database for every phase*, data interchange is still necessary if different groups or companies involved in the design wish to interchange data.

It follows from these considerations that a variety of partners are involved in data intercharge. The sender and receiver of data may be:

- single tools of a toolbox,
- different design teams,
- producers and suppliers (as in automobile engineering), or
- silicon foundries and designers.

3.6.3 Kind and Structure of Interchange Data

Due to the different components which are involved in data interchange and their different use of data, there is considerable variation in the kind and structure of interchange data. These differences can be retrieved in existing formats for data interchange in engineering environments. In the following we want to list some examples and their structure and focus of data.

In engineering applications, formats have been developed to transfer less *structured* data like data for NC-Programming or flat drawing information (Plotfile, GKS-Metafile).

For the needs of interchanging complete product data, formats with complex data structures are in use. Examples are *IGES* for CAD product and geometry data, and *VDAFS* for complex surface description in automobile engineering.

Besides the capability of a format to carry unstructured or structured data, we should consider the *range of data*. According to the purpose of the format, a more or less wide area of engineering applications will be covered. In VLSI design formats exist which cover only parts of a design like *CIF* for VLSI mask layout data. But there are also formats which cover more than one part of a design like *CIDF* for connectivity information and mask layout.

As these examples show, in engineering applications many formats for data interchange have been developed. These formats are essentially tailored to special parts of an engineering environment and are therefore not sufficient for comprehensive engineering applications. In the next section we will outline some requirements for new formats.

3.6.4 Requirements for Data Interchange in New Comprehensive Applications

The development of integrated systems which cover a wide range of applications sets up new requirements for data interchange. The use of existing formats presented above has been limited for many reasons:

- no standard,
- narrow focus,

- interface to other standards,
- proprietary,
- difficult to implement, and
- difficult to extend.

Without a standard many formats for different purposes and different systems exist in parallel. For each format and each system, post- and preprocessors must be written. This results in n*n post- and preprocessors, where n is the number of participants. Whereas, with a standard format only n post- and preprocessors are necessary.

Many formats *with narrow focus* have been developed to address only a specific portion of the total design data problem.

The narrow focus problem might be mitigated somewhat if the new formats were *interfaced to other standard formats* covering other portions of the design data.

Many formats are *proprietary* to specific CAD companies. Their widespread use may be controlled by cost and regulation. The competitive nature of engineering industries usually precludes placing control of such an important aspect in the hands of another single company.

Often, as a format evolves, it becomes increasingly *difficult to develop* a parser or generation software for using it. Issues such as upward compatibility must be considered.

As new design methods evolve, new facets are added to the data that must be transferred. The structure of many formats is such that it is *difficult to extend* them. The usual result is a multitude of enhanced versions which do not match with each other.

3.6.5 Example of a New Format for VLSI Design: EDIF

As stated in the last section, data interchange formats used in integrated systems should cover the whole area of the application data. The format is being used for data interchange between different databases. Therefore, the format must support the exchange of the whole miniworld defined by database schema without any loss of syntax or semantics. To fulfill this requirement, close agreement between format design and database schema design is desirable. A first step in this direction is EDIF – a new approach for interchange formats in VLSI design.

EDIF is targeted at the exchange of design information for gate-array, semi-custom, and full-custom IC design, and for printed circuit board design. This exchange is regarded as a two-way process; i.e., it comprises the route from *foundry to designer* (e.g., macro libraries, technology information), and from *designer to foundry* (e.g., netlist, mask-level geometry, test data).

The structure of EDIF is similar to a nonexecutable, block-structured programming language. EDIF has primitive data like Lisp atoms, and more complex structures built by forming parenthesized lists of primitive data or sublists. The first element of every list is a keyword. Since the semantic emphasis is on keywords rather than syntax, the format can easily be parsed and extended.

The EDIF approach to design is hierarchical. It begins with a *design block*, which identifies a particular cell as the top level of design.

Cells are composed of one or more *views*. A view is simply a specific representation of a cell like mask layout, behaviour, symbolic, schematic, or netlist.

Design hierarchy is established by a cell calling a particular view of a subcell.

To give a feeling for the kind of interchanging data with EDIF, let us now take up our example of a VLSI design. The connectivity of the cells in the example is to be expressed by EDIF Netlist (Fig. 3.35a).

The description of cell CELL1 in view netlist is given below. The cell has input and output ports for interfacing with other cells.

```
(cell CELL1
  (view netlist NET
    (interface
      (define input port (multiple A B))
      (define output port (multiple C D))
    )))
```

In the same manner the cell CELL2 (Fig. 3.35b) can be described:

```
(cell CELL2
  (view netlist NET
    (interface
      (define input port (multiple A B C))
      (define output port (multiple D E F))
    )))
```

Finally, we show the description of CELL3 (Fig. 3.35c). We first describe the interface of CELL3:

```
(cell CELL3
  (view netlist NET
    (interface
      (define input port (multiple A B ))
      (define output port (multiple C))
    )
```

The contents of CELL3 consists of instances of other cells:

```
    (contents
      (instance CELL1 NET I1)
      (instance CELL1 NET I2)
      (instance CELL1 NET I3)
      (instance CELL1 NET I4)
      (instance CELL2 NET I5)
```

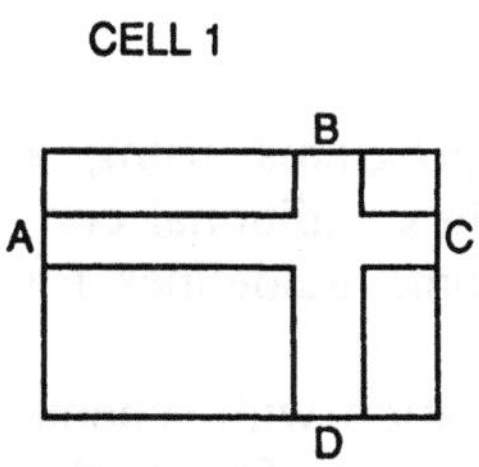

Fig. 3.35a. CELL 1

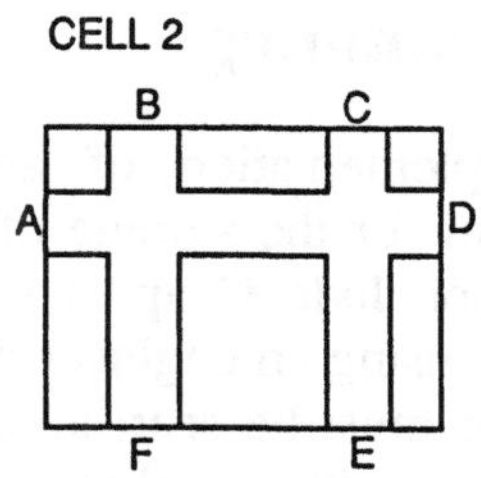

Fig. 3.35b. CELL 2

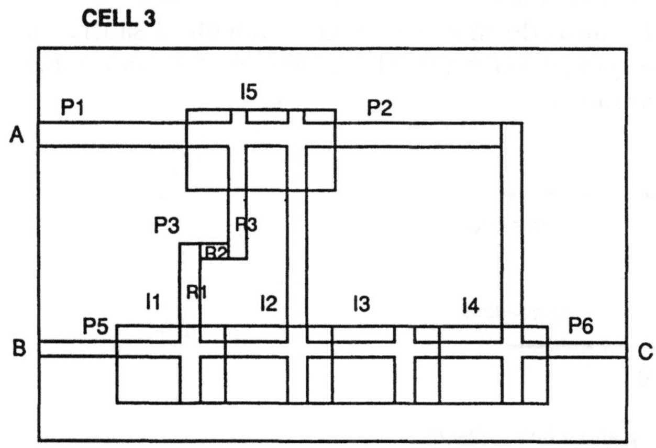

Fig. 3.35c. CELL 3

which are joined together:

(joined A (qualify I5 A))
(joined A (qualify I1 A))
(joined (qualify I4 C) C)
(joined (qualify I5 D) (qualify I4 B)
(joined (qualify I5 E) (qualify I2 B)
(joined (qualify I5 F) (qualify I1 B)
(joined (qualify I1 C) (qualify I2 A)
(joined (qualify I2 C) (qualify I3 A)
(joined (qualify I3 A) (qualify I4 A)

Unused ports are explicitly declared:

(unused
(qualify I5 B)
(qualify I5 B)
(qualify I1 D)
(qualify I2 D)
(qualify I3 D)
(qualify I4 D)
(qualify I3 B)
))))

3.7 Application Programming

The development and implementation of application programs using a database management system for the storage of data requires additional considerations. Therefore,we conclude Chap. 3 by giving some guidelines for writing application programs using an engineering database.

Application programming must be seen in strong correlation with schema design. The applications work with the objects of the miniworld. Due to this fact, they depend on how precisely this world has been modeled by the database schema. From the point of view of application programming, much of the semantics of its special world should be reflected in the database. On the other hand, the modeled miniworld must be general enough to satisfy all applications that are to be integrated via the database. Figure 3.36 shows the conflict between these two positions.

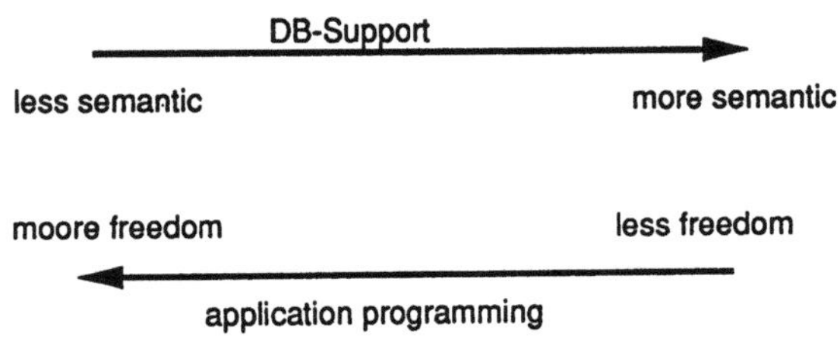

Fig. 3.36. Application programming and schema design

The more semantics is reflected in the database, the easier it is for application programming to store structured data; but also the more specialized will the schema be.

3.7.1 Application Programming in Different Levels of System Integration

The solution to the conflict depends on the degree of system integration. Two typical situations are:

- each application uses its own database, or
- several applications use a common database.

If an application has its *own database*, not used by any other application which needs a fundamentally different schema, the database schema can be designed exactly according to the application requirements. No compromises must be made that would be necessary for a schema for the whole miniworld of all applications. If we have a suitable data model as the basis for the schema design, much of the semantics may be expressed in the database.

In the same way, the interface between the application and the database system can be exactly tailored to the application (see Fig. 3.37).

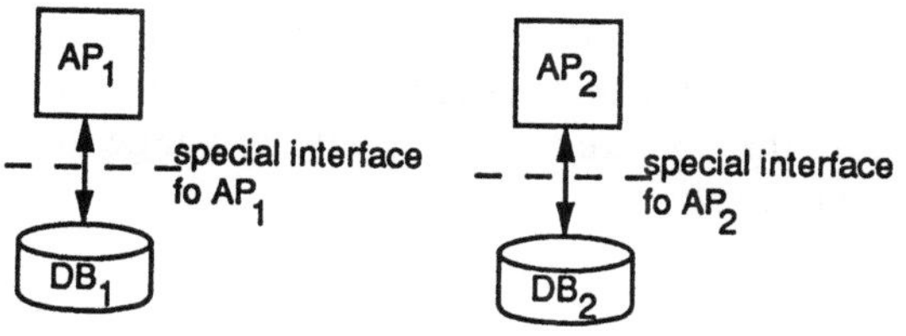

Fig. 3.37. Non-integrated application

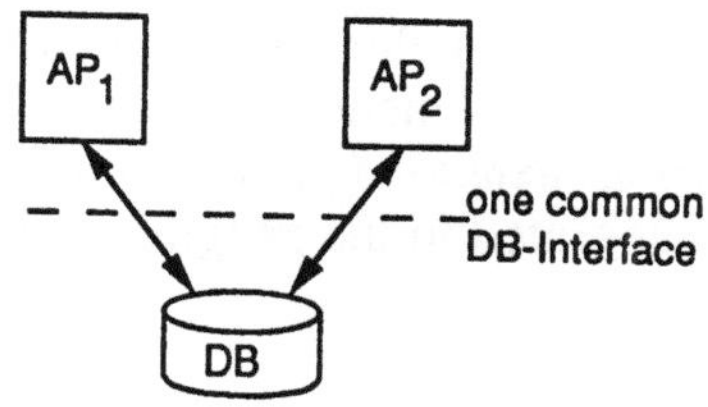

Fig. 3.38. Integrated applications

Integrated applications with a *common database* need a database schema which models the whole miniworld formed by all applications. A compromise must be found between an overloaded, too-complex schema with high semantic content, and poor schemata with most of the semantics hidden in the application programs. Such a compromise may be developed by constructing a semantically rich schema, and then deriving application-specific views or external schemas from it (see Sect. 2.1.2).

Application programs in integrated systems use a common database interface (see Fig. 3.38).

3.7.2 Use of Database Functions for Application Programming

In engineering, the application programs communicate with the database by generic functions, or an operational interface.

If an application programmer uses *generic functions* (functions that are defined independently of a particular schema, as is the case for all data models in Sect. 2.2.2) he needs parameters to refer to the types of a concrete database schema. The database system provides a fixed set of functions whose precise behaviour depends on the schema defined for a special database. To make use of such a function you have to give parameters like object type name or attribute type name. These functions can be used in a call interface, or an extended programming language (see Sect. 2.2.2.1).

A *call interface* provides a set of procedures. Each of these represents a database function, which can be used in a program as follows:

```
.
.
.
statement;
call DBSELECT (p1,p2,...)     <====>    DBMS
statement;
.
.
.
```

Such functions can be embedded easily in application programs, because most programming languages used in engineering environments offer procedural constructs.

Due to the rigid structure of procedural constructs, the formulation of database functions is not as comfortable as in query languages such as SQL. In an *extended programming language*, the constructs of a query language are integrated in the programming language itself:

```
•
•
•
select ... from ....        <====>    DBMS
•
•
•
```

This is a comfortable way of formulating database functions. Since complete integration requires extension of the language definition, and hence construction of a new compiler, a compromise is often sought based on pre-compilers. Transactions P3 and P4 in Fig. 2.51 are examples.

As mentioned above, to use generic functions the application program must have knowledge about the database schema (Fig. 3.39).

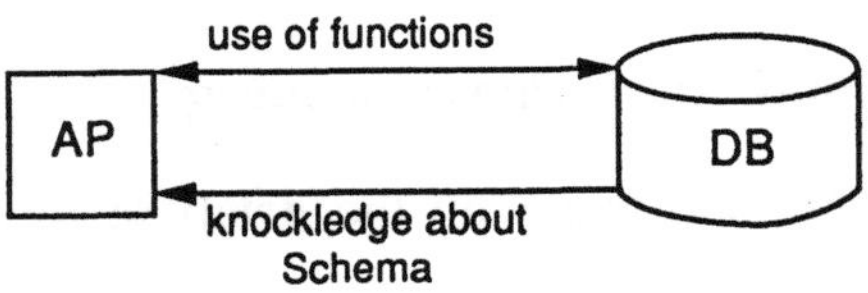

Fig. 3.39. Generic functions

We now present a second class of functions which we will call an *operational interface*. An operational (tool) interface adds a layer between the application programs and the database interface, thus the application does not communicate directly with the database. The new layer projects the functions available for the applications to the database functions (Fig. 3.40). The application program must have little or no knowledge about the schema. The details of the schema are hidden in the operational interface.

A wide-spread method for realizing such an operational interface is the definition of *abstract data types* (ADT). An application works with ADTs and the operations defined for it, which are projected to objects defined by the

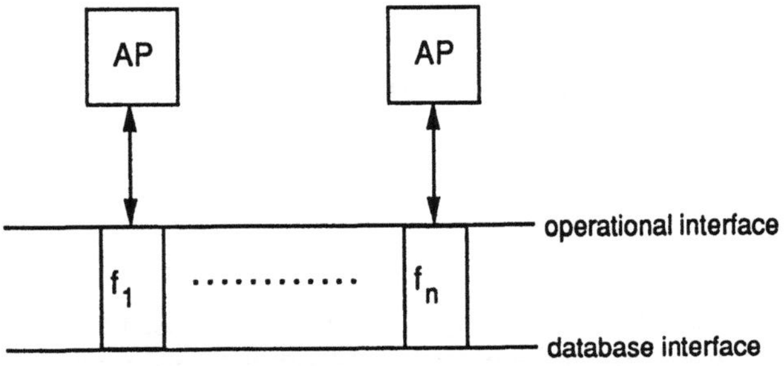

Fig. 3.40. Operational interface

schema and to operations on the database. By building groups of ADTs, the view of a single application to the database can be limited. The application is only allowed to use a subset of the ADTs defined in the operational interface.

We give a small example how abstract data types may be used in electronic design.

Suppose we have a database schema as shown in Fig. 3.41. A complex object "BLOCK" with components "GATE", which can have components "PIN" is defined.

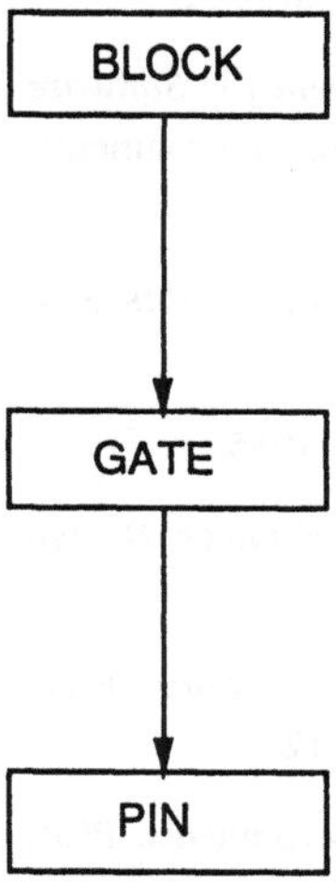

Fig. 3.41. Database schema

Furthermore, the following ADTs are defined (we use a PASCAL-like syntax):

```
function GET_BLOCK (...) : BLOCK;
function GET_GATE (...) : GATE;
function GET_PIN (...) : PIN;
```

If we want to get a specific pin of a gate of a block we can write a program like:

```
•
•
block = GET_BLOCK (...);
gate = GET_GATE (B,...);
pin = GET_PIN (G,...);
•
•
```

Using an operational interface is a comfortable method for handling complex objects by application programs.

3.8 Further Reading

Batz T.: *Versionsverwaltung im Datenhaltungssystem PRODAT des Systementwicklungssystems PROSYT*. To be published in: Proc. GI-Workshop "Datenbanken für Software Engineering", Dortmund 1987 (in German)

Krömker D., Steusloff H., Subel H.-P., (ed.): *PRODIA und PRODAT – Rahmenschnittstellen im Systementwurf*. To be published at Springer Verlag (in German), 1988

Dadam P., Lum V., Werner H.-D.: *Integration of Time Versions into a Relational Database System*. Proc. 10th Int. Conf. on Very Large Databases, Mexico City 1984, pp. 509-522

Dittrich K.R., Gotthard W., Lockemann P.C.: *DAMOKLES – Database System for Software Engineering Environments*. Proc. IFIP Workshop on Advanced Programming Environments, Trondheim 1986, Springer 1986

Dittrich K.R., Lorie R.A.: *Version Support For Engineering Database Systems*. Research Report RJ4769, IBM Research Laboratory San Jose, California 1985

Katz R.H.: *Information Management for Engineering Design*. Springer Verlag 1985

Katz R.H., Lehman T.J.: *Database Support for Versions and Alternatives of Large Design Files*. IEEE Transactions on Software Engineering 10, March 1984, pp. 191-200

Lorie R.A., Plouffe W.: *Complex Objects and Their Use in Design Transactions*. Proc. Database Week: Engineering Design Applications (IEEE), May 1983, pp. 115-121

Riedel-Heine T., Köhler D.: *A Version Management System for Design Environments*. Proc. EUROGRAPHICS 1986, Lisbon 1986, North-Holland Pub. 1986

Tichy W.F.: *RCS – A System for Version Control*. in: Software - Practice and Experience, Vol. 15 (7), July 1985, pp. 637-654

4 Case Studies

4.1 VLSI Design

This chapter shows an actual approach to engineering database systems in the area of the design of integrated circuits. First, an overview is given of the research project E.I.S., sponsored by the German Ministry of Research and Technology, and the situation of a non-commercial open toolbox is outlined. Next, the representation of design information in databases is explained by introducing a data model for design objects and an interface to a relational database system. Finally, an EDIF-oriented database scheme and the adaptation of a tool to this scheme is represented.

4.1.1 The E.I.S. Project – Conception of an Integrated Open VLSI Design System

E.I.S. (Design of integrated circuits) is a cooperative project between German universities, research institutions, and industry. It was started in 1983 by the German Ministry of Research and Technology with the aim of intensifying research and development in the design of integrated circuits, and to spread the science of design methodology.

For this purpose, several goals have been set: Use of uniform hardware and software, collection and adaptation of existing design tools, filling of gaps by new developments, creation of an integrated overall system, and the realization of exemplary applications of electronics. Unlike commercial design software, the university software as used in the E.I.S. project had to satisfy the following conditions:

- available at low cost, quickly and in source code,
- simple to use for educational purposes, and
- operational on a non-workstation environment, i.e., under a time-sharing system with low-cost graphics terminals.

Thus, an open software toolbox resulted from different sources. The universities participating in the project brought in their own development. Some public domain software, mainly from U.S. universities, as well as commercial software is part of the E.I.S. toolbox. Figure 4.1 gives an overview of the design software used in the project.

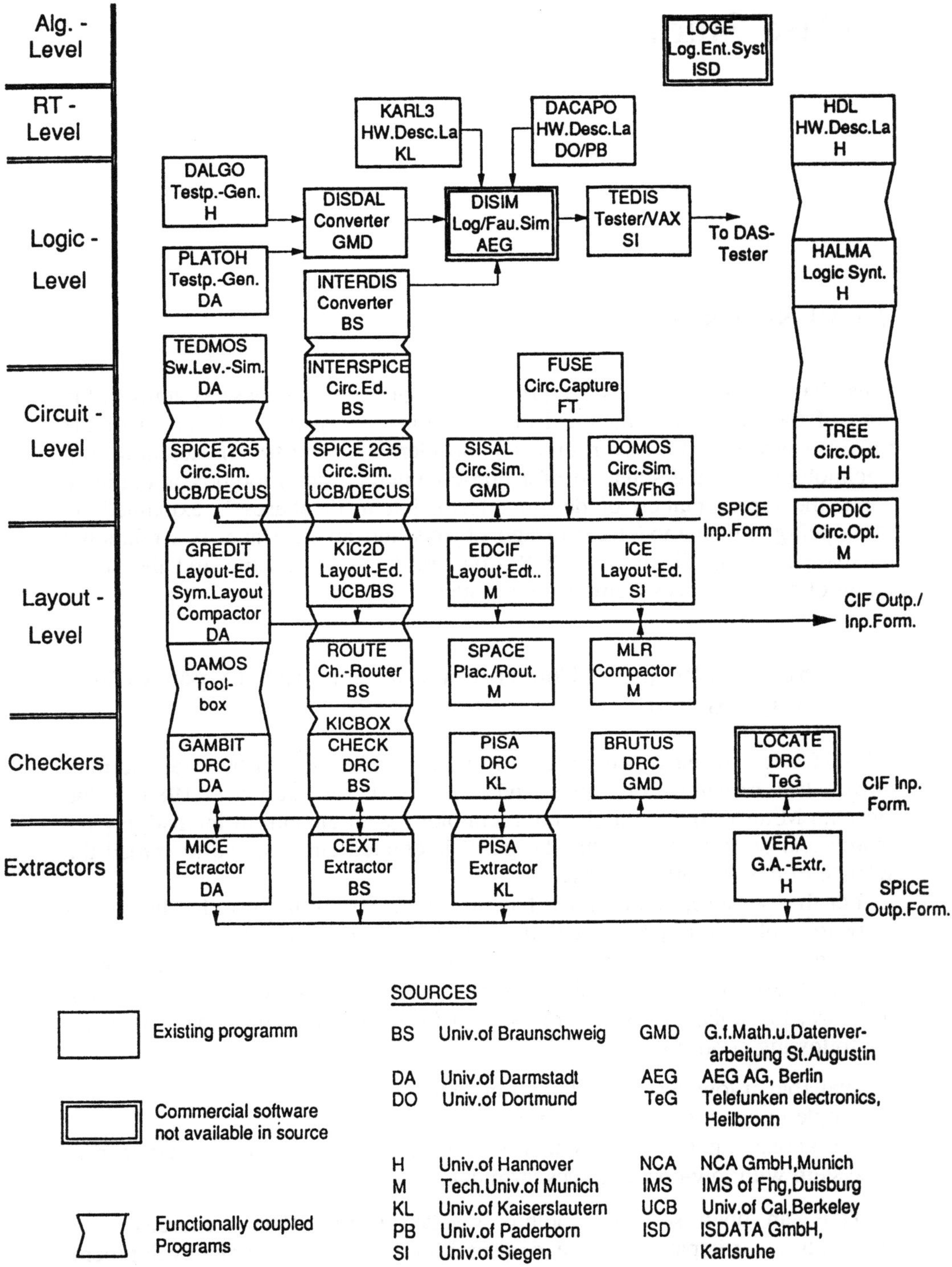

Fig. 4.1. E.I.S. design software

The E.I.S. collection of tools certainly is not going to be an integrated system. The main reason is the lack of a unified database. Therefore, much research is being done in this project into the realization of a VLSI design database system.

4.1.2 Representation of Design Information in Relational Databases

4.1.2.1 Modeling of Design Objects for Extensions of Relational Databases

This section gives an overview about the data modeling concepts in the E.I.S. project. The structural elements for the description of VLSI design data and their mapping to the relational data model are presented.

Complex Objects

An essential characteristic of a database in VLSI design is the heterogeneous structure of the generated objects. Generally, one can assume that each object is composed of several subobjects. The single subobjects in turn frequently consist of subobjects as well. This structure propagates until all branches end in basic objects that are not further decomposable. For this, in applications like VLSI design, the term **complex objects** has been defined.

The following model divides the term complexity into the fundamental requirements and offers to users mechanisms for expressing and managing this complexity in a relational database system. The requirements in VLSI applications use at least two aspects for the complexity of design objects.

First, design objects are represented by highly-structured object types which lead to a heterogeneous but coherent set of simple records or tuples. In

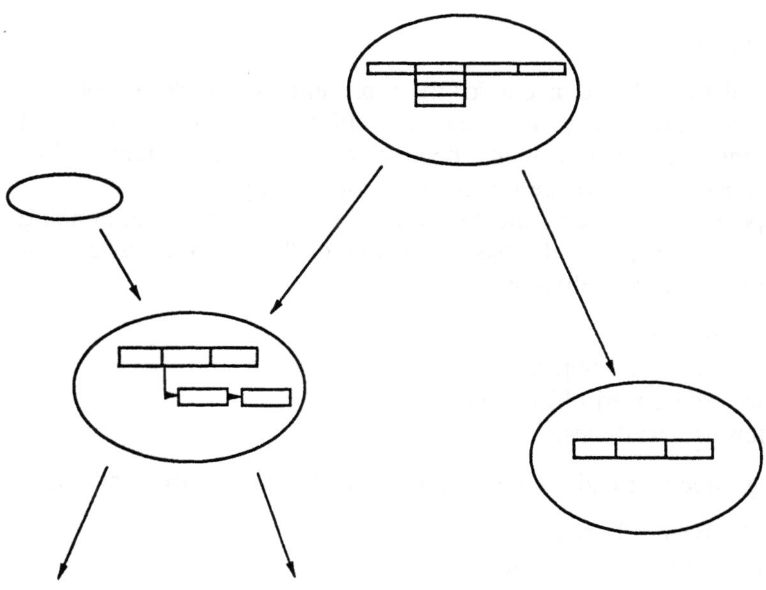

Fig. 4.2. Complex design objects

doing so, the whole set of information is seen as the object of the real world. For this reason it will be the basic unit of storing and manipulating. A dissection into several entities may only be done hidden from the user. The structure of design objects can as well be unsettled sets or lists.

Secondly, objects are described by referencing already existing objects. By using known objects for the construction of complex objects, problems are decomposed into subtasks. With the reference or instantiation relation a complex relationship arises between independent objects of the real world.

The conception of objects used in the E.I.S. model can be described, as in Fig. 4.2. Complex objects are internally of arbitrary structural complexity, and have composition relationships to other objects, which can be concurrently built into other objects.

Highly Structured Objects

The logical view of an entire object is available by the NF^2 data model. See Sect. 2.4.2.3 for further explanation of this model. The structure of a layout object can be defined in a SQL-like language as follows:

```
CREATE TABLE Layout-cell ( Name char,
                          Designer char,
                          SET OF Boxes
                                (x1, x2, y1, y2 number),
                          SET OF Wires
                                (Width number,
                                 LIST OF point
                                      (x-coord, y-coord number)),
                          Labels SET OF char );
```

Composition of Objects

The second part of the E.I.S. concept for the representation of design objects concerns the composition of objects by existing objects. An important aspect was to describe the composition of independent objects in the database. Besides, it must be possible to manage this relation consistently.

For the composition of objects independent of the real world, referencing mechanisms have been added to the basic language of the database system. A reference relation can be described by:

- the name of the reference,
- the type of the referenced object,
- the name of the referenced object, and
- additional reference attributes.

The following three instructions for manipulating references are available:

```
INSERT-REFERENCE named <reference-name>
        from <object-type, object-name>
        to <object-type, object-name>
        set <attribute1 := value, ...>
```

```
DELETE-REFERENCE <reference-name>
        in <object-type, object-name>

CHANGE-REFERENCE <reference-name>
        in <object-type, object-name>
        to <object-type, object-name>
        set <attribute1 := value, ...>
```

The explicit introduction of a reference relation allows the consistency check in complex object hierarchies by the basic data base management system. An essential aspect is the maintenance of referential integrity. This means that objects referenced by an instantiation exist. Furthermore, a normal DELETE of an object must be rejected if some references to this object exist.

But also the user needs some additional mechanisms to restrict or fit the use of references according to his needs. Usually, only references between certain object types are allowed. The definition of objects therefore includes the object types which can be referenced by the defined object. A complete definition of an object generates both the internal and external structure of the object.

```
CREATE TABLE <object-type> (attribute-definition)
    USING {<object-type> (definition of reference attributes)}
```

A complex description of the layout object example can be expressed by a recursive composition relation:

```
CREATE TABLE Layout-cell ( Name char,
                           Designer char,
                           SET OF Boxes
                               (x1, x2, y1, y2 number),
                           SET OF Wires
                               (Width number,
                                LIST OF point
                                    (x-coord, y-coord number)),
                           Labels SET OF char );
    USING layout-cell ( location
                         (x-coord, y-coord number),
                       transformation number)
```

The implementation of the object model on a relational database system can be realized by using the structured objects described above. The reference mechanisms defined by the USING construct can be seen as part of a NF^2 table which cannot be manipulated by the user. This part is only visible for selection and can be manipulated by the above reference functions. A complete presentation of the complex object "Layout_cell" is given in Fig. 4.3.

No additional instructions have to be introduced for the selection of the reference attributes. The USING part can be seen as an extended component of the NF^2 representation (see Fig. 4.3) which can be selected with the usual queries.

Layout_cell

Name	Designer	Boxes				Wires			Labels	References				
										layout cell				
		x1	y1	x2	y2	Width	Points			Ref Na	Obj Na	Attributes		
							x	y				trans	loc x	loc y
2 AND	`xy`	30 5 17	30 5 17	60 70 48	60 70 48	3	5 15	5 30	` text`					
						10	10 20	60 60						
4 AND	`abc`	2 30	2 30	10 40	10 40					A	2 AND		10	10
										B	2 AND		30	10
										C	2 AND		20	60

Fig. 4.3. Representation of the complex object "Layout_cell"

4.1.2.2 Implementing an Operational Interface on Top of the Relational Database System ORACLE

With the intention of getting some realistic impression of the practicability and usability of operational interfaces for highly structured objects, a software prototype was designed and implemented. For this purpose, an operational interface which allows the logical view of highly structured objects has been implemented on top of the commercially available database system ORACLE. The final result is a program shell that supports modeling of complex objects with an extended SQL notation (subsequently called ESQL), which offers additional structuring features like composition and nesting of objects. The aim was to preserve a transparent inclusion of the standard SQL language (both DDL and DML). Internally, all data are still divided into homogeneous subsets, and stored in independent relations.

The actual prototype is realized in a Pascal program which consists of an ORACLE-neutral part and an ORACLE-specific part to be embedded by the ORACLE precompiler. The ORACLE-neutral part basically represents a parser, receiving database instructions in the form of character strings, and reworking them in several steps from left to right. The first steps include SQL-specific preparations (like the handling of lower case, upper case, and special characters, as well as the separation of instructions, text components, protected names, and arguments). During the next process, all language-specific keywords (standard and extended SQL) and valid keyword combinations are extracted and replaced by associated symbols. In this state, the standard SQL words and new ESQL elements are signified differently.

Furthermore, an additional separation of DDL and DML instructions is performed. In an exclusive working unit, each single ESQL statement is converted into an equivalent sequence of standard SQL instructions depending on the new language elements applied. In this context, the accommodation of extended system catalogs to new data objects becomes very important. The individual keywords and the fundamental concept behind the words will be explained subsequently. For the moment, the new keyword RELATION shall be used as a representative to illustrate the whole process. First, the parser checks for the correct use of each recognized keyword in a valid SQL instruction (in the RELATION example it should be the attribute type of a CREATE or SELECT instruction, followed immediately by at least one expression, framed with parentheses, nested to a maximum depth of 2, and containing: name_of_relation [list_of_attribute_names, and definitely no other text]). Secondly, all recognized references are inserted into (or deleted from) the extended system tables (ESYSCAT). Then all of the RELATION constructs (including expressions inside the parentheses) are removed from the original instruction text and replaced by predefined character strings. Thus, the reworked SQL instruction will be added to the previously prepared stack of system catalog instructions and completed by an end_transaction sign. This stack (Pascal array) is passed to the ORACLE-specific part of the ESQL shell where it will be processed sequentially. The ORACLE part of the shell program performs a precompilation of the individual instructions, checks for errors, prepares temporary data buffers (if result data are required), and guarantees a correct termination including the release of locks by an end_transaction operation (COMMIT / ROLLBACK WORK). After successful termination of all ORACLE operations, the temporarily stored result data will be structured and assigned to a predefined Pascal array (tree structure). To achieve a uniform conversion of data coming from ORACLE and a universal handling of the same data in the Pascal environment, a general data type UNIDAT (Pascal record, see Fig. 4.4) has been declared. This data type consists of a variable number of unique atoms. The individual atoms may be pointers, numeric or text data, where data in the form of scalars, vectors, matrices, and tensors are each treated as a closed unit.

```
Type UNIDAT = record
                TSIZ : integer;
                EOFA : integer;
                ATOM : array [0..1000] of record
                          NUL : boolean;
                          TYP : integer;
                          RAN : integer;
                          DIM : array [0..99] of integer;
                          VAD : array [0..1000] of integer;
                          SUB : array [0..99] of integer;
                          end;
                BOOV : array [0..65535] of boolean;
                NUMV : array [0..65535] of real;
                TXTV : varying [65535] of char;
                SPCV : varying [65535] of char;

                EOFB : integer;
                EOFN : integer;
                EOFT : integer;
                EOFS : integer;
                end;
```

Fig. 4.4. General data type UNIDAT

The meaning of each ATOM depends on the value of TYP. If TYP = 0 the ATOM represents a node in the data tree and contains pointers to elements of a substructure or further nodes. If TYP ≠ 0 the ATOM contains elementary data or data references. By means of the component SUB, each ATOM disposes of a variable array of reference pointers which permit nested arrays and the composition of hierarchically structured trees.

Read (SELECT) instructions on highly structured objects will cause nested blocks each containing stacks of standard SQL statements which have to be processed by the interpreter. It can be assumed that in those cases there would be enough potential to improve performance if the existing database system were be replaced by a totally new system, especially designed for this purpose. But such a system would have to be developed completely from scratch, and the effort in designing the database core is far from the purpose of building a prototype or a pilot installation. On the other hand, an optimum of flexibility was reached by the chosen solution of using an existing database system as core, and implementing a software shell as a database extension around this core. This way it was possible to run many experiments with a minimum of effort. Moreover, it is possible to achieve a limited performance improvement even with a usual database system by applying database clustering, and thus storing data physically close which logically belong close together. Surprisingly the response times of the final prototype were quite acceptable, given that the major point of the investigation was to judge the usability and functionality of various semantic and syntactic concepts. For that purpose the following models were considered.

Hiding Storage Structures From the User

This model is not identical to the NF^2 concept, but is based on SQL and represents NF^2 related design objects which can be initiated by "SET OF" clauses as shown in the next example:

```
(1.1)    CREATE TABLE layout_cell (name CHAR(10),
                                   designer CHAR(15),
                                   SET OF boxes (x1 NUMBER,
                                                 y1 NUMBER,
                                                 x2 NUMBER,
                                                 y2 NUMBER))
```

The mapping of statement (1.1) to the internal representation in a relational database is shown together with some illustrating data in Fig. 4.5.

The insertion of data into the table structure will be demonstrated in the next statement:

```
(1.2)    INSERT INTO layout_cell
                             VALUES ('2AND','xy',
                             (30,30,60,60,
                              5, 5,70,70,
                              17,17,48,48))
```

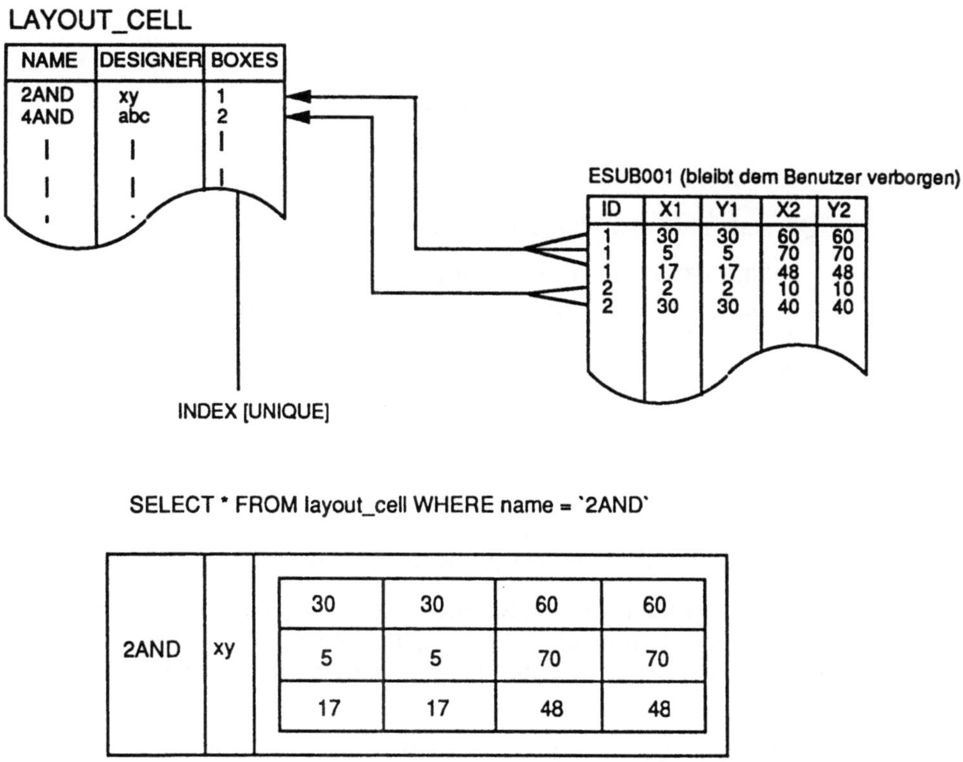

Fig. 4.5. Hidden storage of the "SET OF" structure

The two ESQL statements (1.1) and (1.2) are mapped into standard SQL instructions as demonstrated below:

```
(1.1.1)     INSERT INTO ESYSCAT
                        VALUES (1, 'LAYOUT_CELL', 'BOXES',
                                'ESUB001', '*')
```

The first data value signifies the type of the chosen model (1 = SET OF), the second value contains the name of the source relation, followed by source column(s), target relation (the name ESUB001 will be generated automatically by the program and only used internally), and target column(s). The last value in the INSERT statement is not needed for the "SET OF" type but ESYSCAT will also be used, generally, for the next several suggested types.

```
(1.1.2)     CREATE TABLE LAYOUT_CELL
                        (NAME CHAR(10),
                         DESIGNER CHAR(15),
                         BOXES INTEGER)
(1.1.3)     CREATE UNIQUE INDEX ESUB001I
                        ON LAYOUT_CELL(BOXES)
```

```
(1.1.4)     CREATE TABLE ESUB001 (ID INTEGER,
                                 X1 NUMBER,
                                 Y1 NUMBER,
                                 X2 NUMBER,
                                 Y2 NUMBER)
(1.1.5)     COMMIT WORK
```

The ESQL statement (1.2) translates into:

```
(1.2.1)     INSERT INTO LAYOUT_CELL VALUES ('2AND','xy',1)
(1.2.2)     INSERT INTO ESUB001 VALUES (1,30,30,60,60)
(1.2.3)     INSERT INTO ESUB001 VALUES (1,5,5,70,70)
(1.2.4)     INSERT INTO ESUB001 VALUES (1,17,17,48,48)
(1.2.5)     COMMIT WORK
```

A "DROP TABLE ..." instruction causes the following sequence of internal SQL requests:

```
(1.3)      DROP TABLE layout_cell
(1.3.1)     SELECT * FROM ESYSCAT WHERE TOR='LAYOUT_CELL'
```

TOR is the abbreviation of TO_Relation and represents the target relation. At this point there could possibly be a program interrupt including a warning to the calling instance if the specified table is used as a target (subrelation) in other objects.

```
(1.3.2)     SELECT TOR FROM ESYSCAT WHERE FRR='LAYOUT_CELL'
```

FRR stands for FRom_Relation (= source relation)

```
(1.3.3)     DROP TABLE ESUB001
(1.3.4)     DROP TABLE LAYOUT_CELL
(1.3.5)     COMMIT WORK
```

Visible Hierarchical Structure

The second verified and tested model uses the keyword RELATION to define structured objects and to query existing references. The difference with the SET OF concept is basically determined by the decoupling of superior relations and subrelations, and by the fact that this decoupling is also visible to the user. In particular this means that a user may include subsets of already existing relations as a set of attributes when defining new relations. This concept permits storage and manipulation of data completely independently within the individual relations. Moreover, it is possible to manipulate the intersection with data in the target relation when using WHERE clauses (in usual SQL notation, including ORDER BY, etc.) during data insertion into the source relation. Figure 4.6 illustrates a representation of this model.

The next examples show, according to the present concept, how all affected statement types are converted from ESQL statements into sequences of standard SQL instructions. The first example describes the extended DDL:

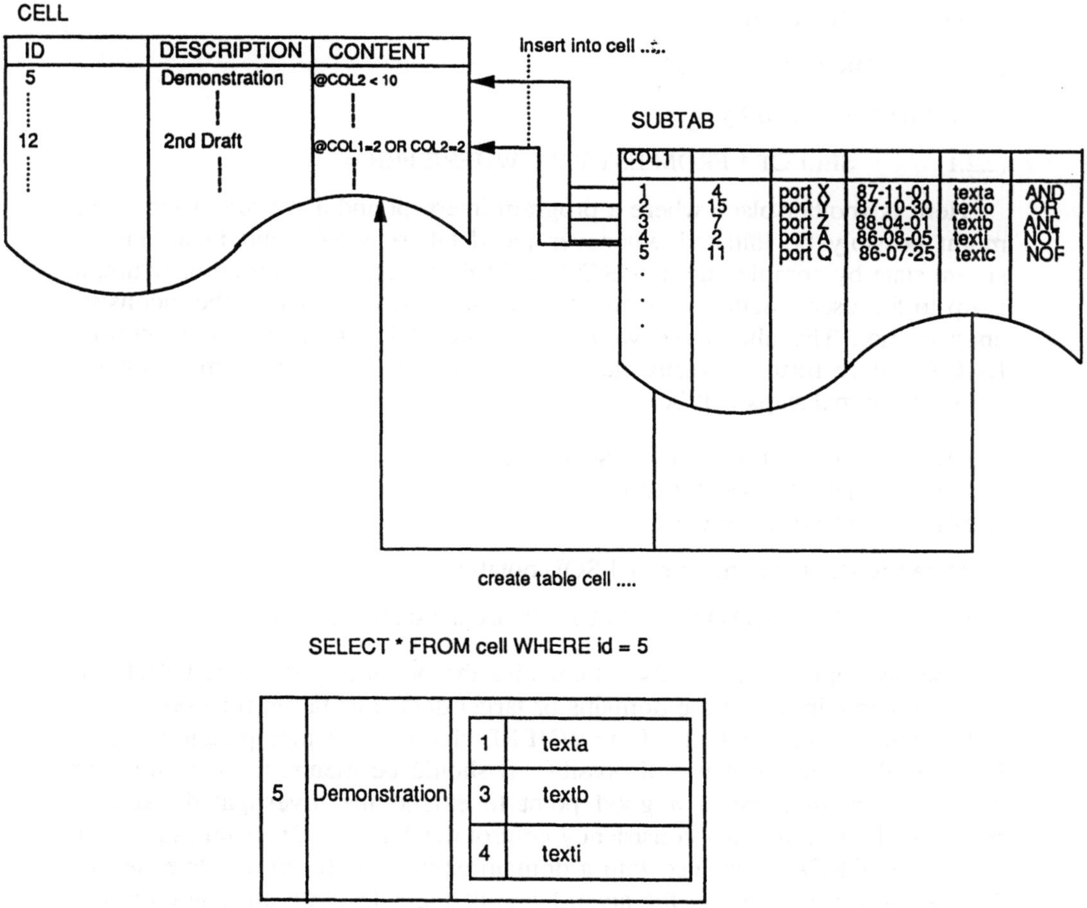

Fig. 4.6. Visible data storage and set combination

```
(2.1)       CREATE TABLE cell (id NUMBER NOT NULL,
                               description CHAR (20),
                               content RELATION (subtab(col1,col5)) )
```

The internal SQL instructions resulting from this statement are:

```
(2.1.1)     INSERT INTO ESYSCAT VALUES (2, 'CELL', 'CONTENT',
                                        'SUBTAB','COL1,COL5')
```

Again the first data value describes the model type ("2" is the identification of type RELATION).

```
(2.1.2)     CREATE TABLE CELL (ID NUMBER NOT NULL,
                               DESCRIPTION CHAR(20),
                               CONTENT CHAR(240))

(2.1.3)     COMMIT WORK
```

The (E)SQL statement

(2.2) DROP TABLE cell

will be performed by:

(2.2.1) SELECT * FROM ESYSCAT WHERE FRR =

Here is another place where a program interrupt and a warning to the user might possibly be initiated whenever the database would end in an inconsistent state by completing the DROP TABLE instruction. After an interrupt, it is up to the user whether he wants to continue and really cause the inconsistency or not. The alternative would be to avoid inconsistency with ROLLBACK and do further investigations on the conflict. If the program continues, the next internal steps will be:

(2.2.2) DELETE FROM ESYSCAT WHERE FRR =
(2.2.3) DROP TABLE CELL
(2.2.4) COMMIT WORK

Data insertion is done in usual SQL notation:

(2.3) INSERT INTO cell VALUES (5,'Demonstration','col2<10')

The last input value (col2<10) contains the previously described WHERE clause for manipulating the domains of target data. The target relation itself is not affected by the contents of the WHERE clause. Later manipulation of data in both tables involved is still possible. It should be mentioned that after the above statement there is a good point to extend the investigated database model with an individual consistency control mechanism. One suggestion is to extend the CREATE syntax with a domain expression (rule) and later let the INSERT operation survey the accordance of the rules with all new data. For reasons of priority, this method has not been realized in the present prototype. The internally deduced stack with standard SQL statements according to the current example contains:

(2.3.1) SELECT * FROM ESYSCAT WHERE FRR = 'CELL'
(2.3.2) SELECT COUNT(*) FROM SUBTAB WHERE COL2 < 10

Statement (2.3.2) may also be used for consistency control.

(2.3.3) INSERT INTO CELL VALUES (5,'Demonstration','@/COL2<10')

The characters @/ are introduced by the program for later recognition of the WHERE clause by the interpreter.

(2.3.4) COMMIT WORK

A standard DELETE instruction only concerns the addressed table and will therefore be taken over without modifications, e.g.,

(2.4) DELETE FROM cell WHERE id = 5

becomes internally:

(2.4.1) DELETE FROM CELL WHERE ID = 5
(2.4.2) COMMIT WORK

In additional to the standard DELETE, there exists an extended DELETE to give the possibility of addressing a whole data structure by one single instruction:

(2.5) DELETE WITH RELATIONS FROM cell WHERE id = 5

Instead of WITH RELATIONS it would also be convenient to use the wording WITH RELATED VALUES. The mapping of this kind of request to the internal interface is represented by:

(2.5.1) SELECT * FROM ESYSCAT WHERE FRR = 'CELL'
(2.5.2) SELECT FRR FROM ESYSCAT WHERE TOR = 'SUBTAB'
AND FRR ~= 'CELL'

The AND clause in the latter statement has been included to protect shared subrelations, which are absolutely valid for the present concept.

(2.5.3) SELECT CONTENT FROM CELL WHERE ID = 5

The previous statements as well as the following ones have to be executed occasionally in nested loops with continuous modifications depending on the current table contents and on the defined data structure.

(2.5.4) DELETE FROM SUBTAB WHERE COL2 < 10
(2.5.5) DELETE FROM CELL WHERE ID = 5
(2.5.6) COMMIT WORK

It should be indicated that the submitted models do not allow the use of highly structured data units as operands within WHERE clauses. The technical realization of such a service as well as the investigation of possible effects to the entire database language are both extensive research areas which will not be considered in this context.

When manipulating subordinate data through a superior relation there should be a distinction between direct data manipulation like inserting data values, and indirect data manipulation like modifying join conditions. The next example shows the scheme of indirect data manipulation:

(2.6) UPDATE cell SET description = 'New text',
RELATION(content) = 'col1=77' WHERE id = 5

Mapping:

(2.6.1) SELECT * FROM ESYSCAT WHERE FRR = 'CELL'

Statement (2.6.1) is needed for consistency control.

(2.6.2) UPDATE CELL SET DESCRIPTION = 'New text',
CONTENT = '@/COL1=77' WHERE ID = 5
(2.6.3) COMMIT WORK

The direct data manipulation through a source relation has not been implemented in the present prototype (a separate manipulation of subordinate and superior relations, however, is easy and quite acceptable), but a possible solution is presented in the following example:

(2.7) UPDATE cell SET description = 'New text',
content = (26,'texta',28,'textb',.....) WHERE id = 5

Proposal:

(2.7.1) SELECT * FROM ESYSCAT WHERE FRR = 'CELL'

After this operation there must be a consistency check which should cause an interrupt when the modified data overlap with the contents of further relations (sharing of subrelations). For this case, the user can be given a chance to force the UPDATE by introducing a new keyword (e.g., REPLACE or FORCE):

UPDATE .... content = (REPLACE:26,.......)

(2.7.2) DELETE FROM SUBTAB WHERE COL1 = 77
(2.7.3) INSERT INTO SUBTAB(COL1,COL5) VALUES (:1,:2)

There is, again, a high risk of inconsistencies at this point, if at least one of the attribute types referenced in the active WHERE clause is not simultaneously an element of the attribute type definition in RELATION(subtab). These possible inconsistencies can be removed either interactively by requesting and adding the missing relative values during the UPDATE operation, or in general by adapting the join conditions. Furthermore, all attribute definitions which are not referenced must accept SQL NULL values. Operation (2.7) terminates with the internal statements:

(2.7.4) UPDATE CELL SET DESCRIPTION = 'New text' WHERE ID = 5
(2.7.5) COMMIT WORK

The last group of SQL statements affected by the investigated model are the SELECT statements. The first example from this group will show the internal mapping of a standard SELECT statement:

(2.8) SELECT * FROM cell

According to the previous example data this operation is decoded into:

(2.8.1) SELECT * FROM ESYSCAT WHERE FRR = 'CELL'
(2.8.2) SELECT * FROM CELL
(2.8.3) SELECT COL1,COL5 FROM SUBTAB WHERE COL1 = 77

It is obvious that in the Pascal shell the individual SELECT statements will be decoded into the basic SQL units PREPARE, DESCRIBE, OPEN, FETCH, and CLOSE.

PREPARE permits a precompilation of instructions and recognizes semantic and syntactic errors. In case of frequently repeated SQL statements this preparation can lead to remarkable time advantages. For further time optimization, the most frequently reused statements like SELECT * FROM ESYSCAT WHERE FRR = '....' have been extracted from the dynamic

ORACLE routines and put into individual procedures, prepared and optimized for static SQL. These procedures are finally bound to ORACLE and compiled with the entire Pascal program.

DESCRIBE is needed to obtain job-relevant information such as data types, pointers, length values, buffer size, etc.

The remaining units OPEN, FETCH, and CLOSE are necessary to realize nested read instructions like (2.8.2) and (2.8.3).

A more specific SELECT statement is demonstrated next:

(2.9) SELECT content FROM cell WHERE id = 4

The internal mapping to this statement will be:

(2.9.1) SELECT * FROM ESYSCAT WHERE FRR = 'CELL'
(2.9.2) SELECT CONTENT FROM CELL WHERE ID = 4
(2.9.3) SELECT COL1,COL5 FROM SUBTAB WHERE :execute_content

If statement (2.9.2) returns (as in the current example) an empty result, statement (2.9.3) will be skipped. Another new feature of the extended SQL model is the use of the keyword RELATION in SELECT expressions to receive the description of selection criteria instead of the referenced data themselves. The semantic representation of such an example is:

(2.10) SELECT RELATION(content), description FROM cell

The mapping to standard SQL delivers:

(2.10.1) SELECT * FROM ESYSCAT WHERE FRR = 'CELL'

This statement is mainly needed to guarantee the valid use of the word RELATION.

(2.10.2) SELECT CONTENT, DESCRIPTION FROM CELL

The direct result of column CONTENT (e.g., '@/COL2 BETWEEN 20 AND 45') might be reworked together with the descriptive values coming from query (2.10.1) into the final output of a contents description of the referenced structure in typical SQL notation. According to the above example the result would be: COL1,COL5 FROM CELL WHERE COL2 BETWEEN 20 AND 45.

4.1.3 EDIF-Oriented Design Systems

Two steps are necessary if the DBMS extension will be used in a VLSI design environment. On the one hand, a database schema, tailored to the application and the tools must be designed. On the other hand, an interface for tools is required, which allows effective object-oriented access to the tool design data.

Whereas no common models exist for the representation of the "VLSI world" in databases, several formats for the exchange of design data between different CAD systems exist at present. These formats describe the design data in a uniform way, which is syntactically and semantically definite for both the sender and the receiver. The "Electronic Design Interchange Format (EDIF)" has become the most important format among data exchange standards for electronic design.

Although EDIF was planned only to be a pure data interchange format, there are good reasons to use EDIF as the basis for a database scheme as well:

- EDIF offers a uniform, standardized view to the bulk of data arising in the design cycle.
- Some of the object structures needed for VLSI data models are already present in EDIF.
- The orientation toward a generally accepted transfer standard simplifies the exchange of data between different CAD systems.
- Tools which already have an EDIF-oriented internal data structure can be easily adapted to the database scheme.

4.1.3.1 EDIF-Oriented Database Schemes

In this section we give an example of an EDIF-oriented scheme based on the data model presented in Sect. 4.1.2. The scheme contains the ports and the graphical symbol of a VLSI cell. Furthermore, the composition of the cell from instances of other cells and nets for connecting the ports are modeled.

Figure 4.7a shows an extract of an EDIF file describing a 1-bit full adder, which is composed of two half-adders and one OR gate. Figure 4.7b contains the schema definition for such an object.

4.1.3.2 The EDIF System – Support Tools for Checking and Previewing EDIF

Now we give an example of how to adapt a tool to the database system. For the storage of data received by an EDIF file, a tool must be available which parses the file and inserts the correct data into the database. With an intermediate step – the use of an internal data structure – an efficient access to the database is possible. The units of insertion are not single values like rectangles or points, but complex objects like ports or complete symbols.

The development of EDIF-related software makes tools for the validation of EDIF files and for graphical preview of geometric data indispensable. With the EDIF system, full syntactic and semantic checking of EDIF files is possible. Furthermore, the graphic representation of EDIF data by GKS is supported.

Processing EDIF Files – The Parser

The EDIF Parser accepts an EDIF file as input and generates an internal data structure. The input file is processed in four steps: Scanning the input, syntax analysis, construction of the internal data structure, and semantic analysis.

A complex data structure containing all information for further processing of EDIF data is built during the parsing process. The generated "EDIF tree" is syntactically absolutely correct.

```
(cell FULL_ADDER
  (cellType GENERIC)
  (view LOGIC
    (viewType SCHEMATIC)
    (interface
      (port X (direction INPUT))
      (port Y (direction INPUT))
      (port CIN (direction INPUT))
      (port Z (direction OUTPUT))
      (port COUT (direction OUTPUT))
      (symbol
        (figure SG
          (rectangle (pt 10 10) (pt 125 80))
          (path (pointList (pt 0 60) (pt 1 60) ))
          ......
        )
        (boundingBox
          (rectangle (pt 0 0) (pt 135 90))
    ) ) )
    (contents
      (page ADDER
        (instance I1
          (viewRef LOGIC (cellRef HALF_ADDER))
          (transform (origin (pt 25 35)))
        )
        .......
        (net A
          (joined
            (portRef Z (instanceRef I1))
            (portRef Y (instanceRef I2))
          )
          (figure ..... )
)))))
```

Fig. 4.7a. EDIF file

```
CREATE TABLE Schematic_cell (
    Name char,
    Viewname char,
    SET OF Port (
        Name char,
        Direction char
        ),
    Symbol (
        SET OF Figure (
            Group char,
            SET OF Rectangle (
                X1, Y1, X2, Y2 number),
            SET OF Path (
                LIST OF Point (
                    X, Y number))),
        BoundingBox (
            X1, Y1, X2, Y2 number)),
    SET OF Page (
        Name char,
        SET OF Net (
            Name char,
            Joined (
                SET OF (
                    Port char,
                    instance char
                    )),
            SET OF Figure ( ..... )
    ) ) )
USING Schematic_cell ( transform (
                    origin ( X, Y number ))
            )
```

Fig. 4.7b. Schema definition

The internal data structure is implemented as an abstract data type (ADT). This means that the internal data structure is hidden to the applications working with the structure. The parsing algorithms themselves use the ADT for building the internal structure. The manipulation of, and access to the ADT "EDIF tree" takes place only by algebraically specified functions. The semantics of the functions are transparent, as they are fully described by the axioms of the data type. With nodes of type KEYWORD, SYMBOLIC_CONSTANT, IDENTIFIER, INTEGER, STRING, and POINTER, the tree reflects the EDIF hierarchy. Depending on the node type, different attributes are associated with each node. Whereas KEYWORD nodes do not have any attribute, an INTEGER node, for example, contains the integer value, and a POINTER node the pointer to the referenced tree. The parser only accesses the internal data structure which is mapped to the database. Figure 4.8 shows an example for the EDIF tree. We use a graphic representation to display the top level of the data structure.

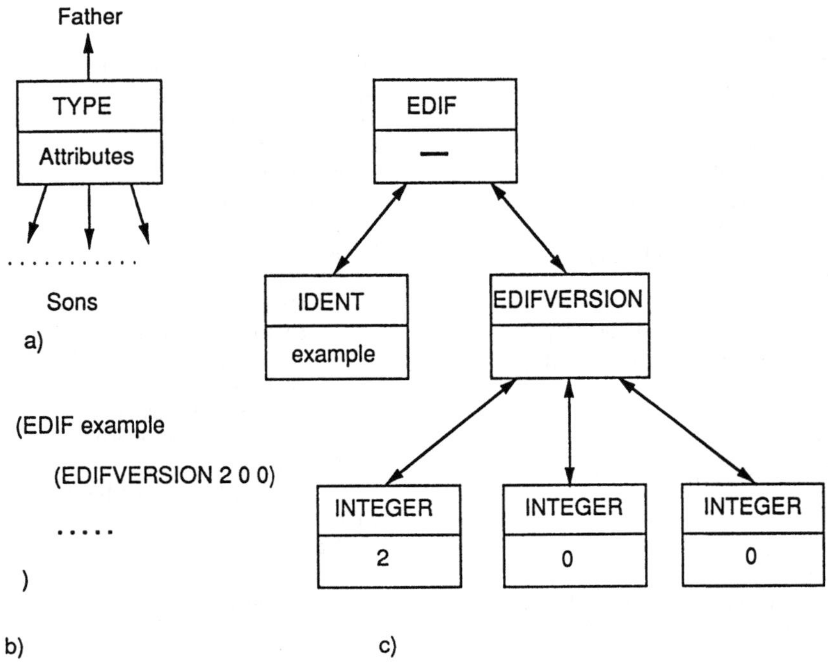

Fig. 4.8. Internal data structure

The functions of the ADT are also used for accessing the design data. The data are stored in a database system, whereas the internal data structure allows easy and efficient access by tools.

The following functions for tree construction exist:

- **MAKETREE : KEYWORD -> TREE**
  Generates a new node and registers the keyword.
- **ENTERID : TREE x IDENT -> TREE**
  Stores the identifier in the root of the tree, if the tree is generated as Identifier, otherwise error.
- **ENTERINT : TREE x INTEGER -> TREE**
  Stores the integer value in the root of the tree, if the tree is generated as Integer, otherwise error.
- **ENTERSTR : TREE x TSTRING -> TREE**
  Stores the string in the root of the tree, if the tree is generated as String, otherwise error.
- **ENTERPOINT : TREE x TREE -> TREE**
  Stores a pointer to the second tree in the root of the first, if the tree is generated as Pointer, otherwise error.
- **ADDSON : TREE x TREE -> TREE**
  Adds the second tree as last son of the first one.

Access to the tree can be made with the following functions:

- **GETKW : TREE -> KEYWORD**
  Returns the keyword of the root node.
- **GETID : TREE -> IDENT**
  Returns the identifier of an identifier node.
- **GETINT : TREE -> INTEGER**
  Returns the integer value of an integer node.
- **GETREAL : TREE -> REAL**
  Returns the real value of an node of type "(**e** *integer integer*)".
- **GETSTR : TREE -> TSTRING**
  Returns the text string of an string node.
- **GETPOINT : TREE -> TREE**
  Returns the root node of the referenced tree.
- **GETSON : TREE X INTEGER -> TREE**
  Returns the i-th son of the tree.
- **GETFATH : TREE -> TREE**
  Returns the father node, if it exists.
- **NUMBSONS : TREE -> INTEGER**
  Returns the number of sons.
- **CUTSON : TREE -> TREE X TREE**
  Cuts the last son of the tree and returns it as a separate tree.

Context Analysis

The context analysis checks the semantic correctness of an EDIF file and covers the following points: Recognition of name conflicts, declaration of names, referencing, use of context dependent constructs, dimensions, and use of keyword display. This analysis works on the internal data structure and accesses the database to get the information.

The semantic checks are embedded in the syntax analysis: Context tests are performed after the identification of the corresponding syntactic construct. For example, the uniqueness of a library name is checked directly after parsing the LIBRARY construct.

Previewing EDIF – The GKS Interpreter

The GKS preview utility allows the visualization of graphical EDIF data on a GKS workstation. Data structures and interchange formats for product definition data usually distinguish between geometric / topologic information and representation information. Geometry and topology are the typical areas for CAD systems; representation is the task of graphic systems like GKS. However, the GKS interpreter can only visualize basic graphics obviously contained in an EDIF file. Processing of complex geometry will remain the task of a sophisticated CAD system. Besides, the main purpose of this utility is exclusively to offer a tool for getting a quick overview of an EDIF file.

Accessing The Internal Data Structure

The GKS interpreter works on the internal data structure. General information about an EDIF file can be retrieved via an interactive dialogue:

- a list of designs,
- a list of libraries,
- a list of cells, and
- a list of views.

The desired information is retrieved from the database. Furthermore, single views can be selected for the display of graphical information. The user may select the depth of hierarchy and can choose between the representation of the interface or the contents section of a cell. Then the appropriate part of the design is selected from the database and the internal data structure is built.

The graphical data in the selected part of the internal data structure are displayed using GKS functions. The interpreter starts at the root of the subtree and navigates through the tree in a recursive descending manner, whereby all nodes without a meaning for the graphic representation are ignored.

4.2 Software Engineering

This section first takes a look into the field of application software and system engineering (SE) by glancing at the framework an engineering database system could be embedded in. After that, the PRODAT data model of the SE environment PROSYT is discussed, wherein simple and structured objects (our notion of complex objects) are introduced. Afterwards, a suggestion of mapping the application-oriented database model of PRODAT to the kernel system DAMOKLES for engineering databases follows. This case study mainly considers the basic mechanisms for handling complex objects. The traditional data models mostly suffer from lack of semantics (e.g., the relationships in relational models can only be identified by interpreting attributes) for non-standard applications. The property *semantics-oriented* means that the mechanisms used o represent the structure of the managed data should correspond as closely as possible to the meaning that database users associate with the data. Therefore, the last subsections of Sect. 4.2 focus on the special relationships between parts of complex objects, which can be described by adding certain consistency constraints for objects and relationships. This section should cast light on the usage of PRODAT and DAMOKLES in a special application area.

4.2.1 Software Engineering Environments

4.2.1.1 Requirements of Software Engineering

The complexity of software and system design process control requires computer support for all phases of the system life cycle covering specification, design, implementation, maintenance, project management, and quality control. We have analyzed some approaches and projects, which will determine new products on the future market. Certain engineering environments with the goal of a complete integrated software system (including the engineering database) are currently under development. Let us first summarize some general requirements for such systems.

Existing software systems are mostly not standardized products, with the risk of errors, updates, and incompatible versions for users. The main objective of current software engineering projects is the development of a complete system by integrating existing and new design tools. Integration of tools and tool sets requires an open system environment with well defined interfaces.

Users need combinations of tools, or a choice of tools out of tool sets; this is not realized by the addition of single computer based tools.

The consistent use of compatible and complete tools in a common environment also means the solution of problems with a certain kind of philosophy for software production.

In some applications, complexity of systems requires the inclusion of hardware design to reach performance control. Such systems enclose computer-based tools supporting all phases from requirements analysis, through specification and tests, to maintenance of products. Furthermore, work in project management, documentation, and quality control has to be supported in the same environment.

Acceptance of a software engineering product also depends on:

- portability,
- flexibility, and
- extendability.

The use of certain software tools is not fixed to special hardware. Portability can be archieved by using UNIX as the basic operating system. Changing and combination of different tools in the software environment is possible, which guarantees a flexible use of components. New methods need new tools, and users will want to extend their systems by integrating these new tools.

The basic concept of current projects in this area is the integration of tools by adapting them to standardized interfaces – an interface to a common database system and to a uniform user interface. Both interfaces are named the frame of the system. Hence the components of the frame are object management, tool communication, and man-machine dialogue. In Sect. 4.2.1.2 we describe an example of such a frame to understand the use and environment of an engineering database system in software production.

The database system is the base, which supports object management and tool communication. Usual design tools use the file management of the operating system directly, but this limits portability. Some tools use relational database systems; we have seen their limits and the necessity of new data models in the previous chapters. The user interface described supports features like window management, mask and menu handling, and graphics. These uniform user interfaces have to use the database interface also, and therefore they may have special requirements of the database component.

In summary, the main requirements for a common database system in Software Engineering are:

- modeling of object interrelations by a predefined set of frequently needed relations,
- design of objects using predefined objects,
- an integrated archiving system,
- version and configuration control, and
- management of long transactions.

How the database components try to meet these goals is explained in Sect. 4.2.2, where the features of PRODAT are described as a non-standard database system for system engineering.

4.2.1.2 A System Engineering Environment

PROSYT* is a system engineering environment intended to support the developer of complete computer systems (this means not only software, but also hardware and firmware) through the whole system life cycle from requirement specification to maintenance. As there exists no standard tool covering the whole system life cycle, it is necessary to combine a number of tools to form a toolkit. The goals of PROSYT are to use existing tools, and to provide an open environment with the ability to integrate future tools easily. This is done by a system architecture shown in Fig. 4.9.

The PROSYT frame consists of the four parts PRODIA, PRODAT, POE, and Project Advisor: PRODIA, the user interface manager which presents the

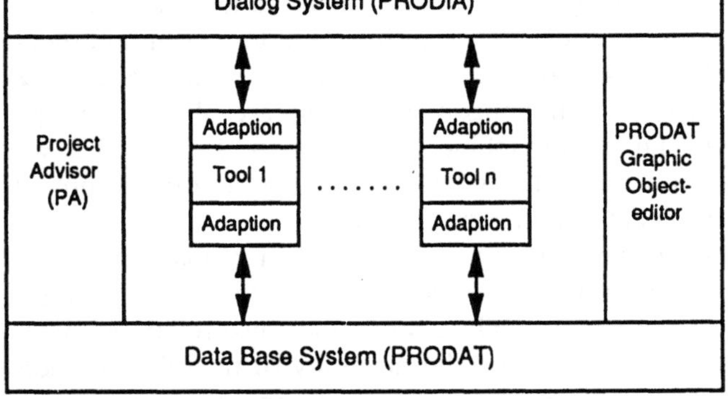

Fig. 4.9. PROSYT system architecture

* PROSYT is a research project sponsored by the Department of Research and Technology of the Federal Republic of Germany under Contract No. ITS8306A7.

user with homogeneous behaviour of the different tools; the Project Advisor, a knowledge-based tool to retrieve reusable parts of previous projects; the common database system PRODAT; and the PRODAT Object Editor, POE as a graphics tool for inspecting and manipulating the objects of the database interactively.

Inside this framework, several toolkits are embedded. There is more than one toolkit, because users need different configurations of tools. Frequently there are tools supporting the same design phase, but employing different methodologies.

The elements of a toolkit have mostly been developed as stand-alone tools, with their own data management and user interface. Combining them in a homogeneous way means adapting their user interface to a more common interface which supports users working with several tools (by methods like multi-windowing, etc.). Besides, tools must be enabled to work with the results of other tools.

The last topic – the exchange of data – will now be dealt with in more detail. There are three possible ways of exchanging data. First, one could allow tools to access data of other tools directly, but then each tool has to know the internal data structures used by all other tools and could not be kept from accessing them at all times in whatever internal status they may be. The second solution is to build data transformers between all tools working directly together. Two main problems arise. Such a system is not open in the sense that it is possible to integrate new tools, and the number of transformers increases enormously with the number of tools. A central problem in both solutions is the lack of a facility to store data concerning more than one phase of the system life cycle (i.e., pertaining to more than one tool), or to the whole project associated with no specific tool.

The third solution – chosen in PROSYT – is connection through a common database system. Each tool must be adapted just once to that database system, so the number of transformations decreases. The system is open to integrate additional tools easily, by building only one transformer to the database system. Data concerning the whole project are, just like data from the several tools, stored in one place (the database system), and consistency, access rights, and versioning can be supported.

4.2.2 The PRODAT Object Model

In this section, an informal overview of the PRODAT data model is presented. System engineering needs a data model that allows a natural (intuitively direct) and flexible definition and evolution of highly structured objects, which are composed of other objects. A point that has been accepted in this area is the view that an object can be described by a certain contents, a number of properties or attributes, and a structure. The structure is the composition of parts to represent a whole entity. What *composition* or *is_part_of* means is determined by the application. The application requires additional semantics here, which can be derived from the database system through the schema.

Another requirement is that it be possible for objects to share subobjects to avoid strong hierarchies. Recursion should be supported by the model when defining objects. The fact should not be hidden that software modules, e.g., are constructed from other modules. Additionally, the system has to support versions in a way which does not force an explicit modeling, but a user-controlled version management.

Therefore, PRODAT provides mechanisms for modeling complex objects and design interdependencies, which include basic operations for:

- construction of complex (structured) objects from predefined parts or subobjects,
- recognition of the construction of a certain instance of a complex object, and
- selection of special parts of complex objects.

There are other concepts based on the object model with structured objects which are not described here. It is necessary, of course, to have an adapted transaction mechanism with locking units of the object model, and with access rights with a fine granularity (for objects and functions). Another concept supported is the distinction between public and working databases. In working databases the private use of objects, which are transferred from public databases by a checkout mechanism and special locking modes, should be guaranteed. Archiving of objects for a long period of time without leaving the database system is a requirement concerning design and project management. PRODAT will meet this goal by an integrated archiving component, where archiving of objects, versions, and configurations is done in a user-controlled fashion. The concepts were introduced by some examples in Sect. 3.5.

4.2.2.1 Simple Objects

The basic entities of the PRODAT object model are *simple objects*. According to their type definition, they may have a so-called *contents*, and *attributes*. The contents are a variable length string of bytes which is not interpreted by PRODAT. For example, the contents of a simple object can be text, object code, or pixel data. Attributes describe characteristics of objects. The user may define his own attributes based on several primitives (e.g., CHAR, INTEGER, or BOOLEAN). Attributes are similiar to those of the relational model (see Fig. 4.10); each attribute has a basic domain type and is not structured (1Nf, compare with Sect. 2.4.1). Examples of attributes are:

- creation date of an object,
- name of author,
- geometric data, and
- status information.

Each PRODAT object has a system-generated internal identifier which is a unique key within the system. Additionally, objects may have an arbitrary user-supplied name.

Attribute	Domain
01.08.87	date
heiner	NAME
NAME	char 15
160559	integer
Contents	
program gen_pic begin ... end.	

Fig. 4.10. Attributes of an object

4.2.2.2 Structured Objects and Relationships

Structured objects are built from the simple objects introduced above by means of relationships. Relationships are binary and directed. Objects used in another object via relationships are called the subobjects or successors of that object. A *structured object* is defined as a simple object – its root – and a (possibly empty) set of subobjects whose members are connected to the root by relationships; the subobjects may be structured themselves. An object may be used in arbitrarily many structured objects ("shared object"). In other words, the resulting structure does not form a strong hierarchy as in the NF^2 data model.

The recursive nature of a structured object is reflected in its type definition, which consists of two parts. First, the simple object, which is its root, is described. Second, the type and number of admissible subobjects is indicated in the SUCCESSORS clause.

The example in Fig. 4.11 shows an illustration. Objects of type *module* should have one subobject of type *interface*, and an arbitrary number of subobjects of type *implementation*. In the object instance shown in the picture, there is a structured object *finset* with one interface subobject *finset.int*, and two different implementation subobjects, *array.imp* and *list.imp*. Thus the structured object *finset* complies with its type definition.

```
OBJECT TYPE  module
             (      ...
                    SUCCESSORS 1 interface AND 1..* implementation
             );
```

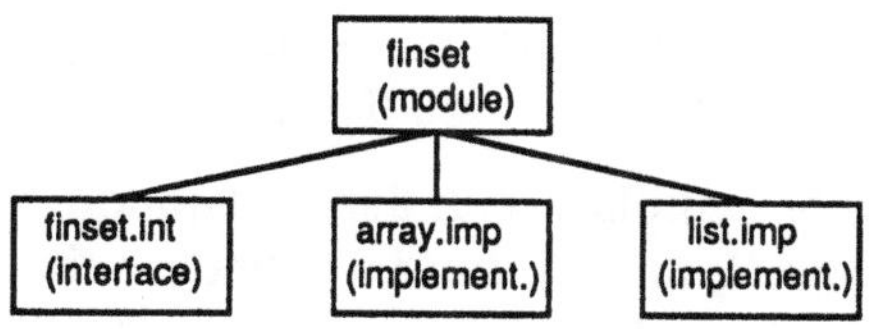

Fig. 4.11. Definition and instance of a structured object

In order to explain the SUCCESSORS clause, we introduce a graphical representation. Let a *constellation* be a simple object together with the set of all its immediate simple successor objects. The graphical representation of a con-

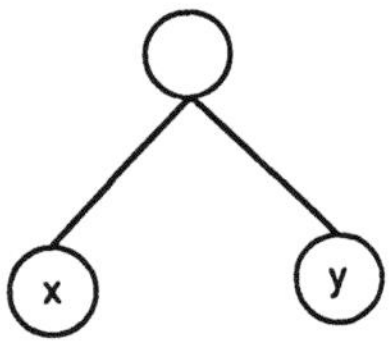

Fig. 4.12. Constellation of an object o with subobjects x and y

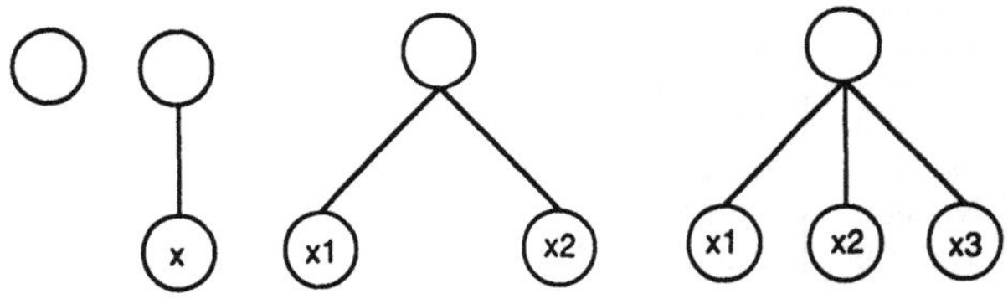

Fig. 4.13. Constellations of SUCCESSORS X

stellation is a tree-like figure with depth one (Fig. 4.12), whose leaves are marked with the successor instances. We represent these by small letters corresponding to the (capitalized) type names. A subscript helps to distinguish if there is more than one object of the same type (e.g., two objects of type X are named x_1 and x_2, respectively).

The SUCCESSORS part of the type definition describes the admissible constellations of an object of this type. In the simplest case there is only one type name given, e.g., *SUCCESSORS X*. Then, arbitrarily many subobjects of type X are allowed (Fig. 4.13), including the case "no successors".

Type names may be prefixed with a cardinality declaration. This can be a number, an interval, or a list of numbers and intervals. The symbol "*" appearing as the upper bound of an interval means "infinity". For instance, the semantic of *SUCCESSORS 1, 3..* X* is given by "either one object of type X or at least three of type X". Figure 4.14 shows the constellations specified by this statement. *SUCCESSORS X* is by default equivalent to *SUCCESSORS 0..* X*.

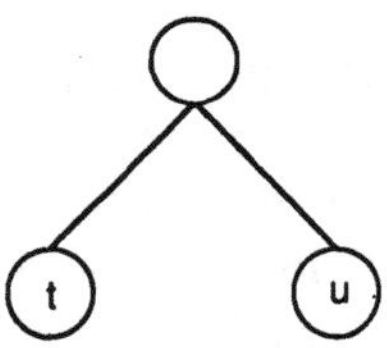

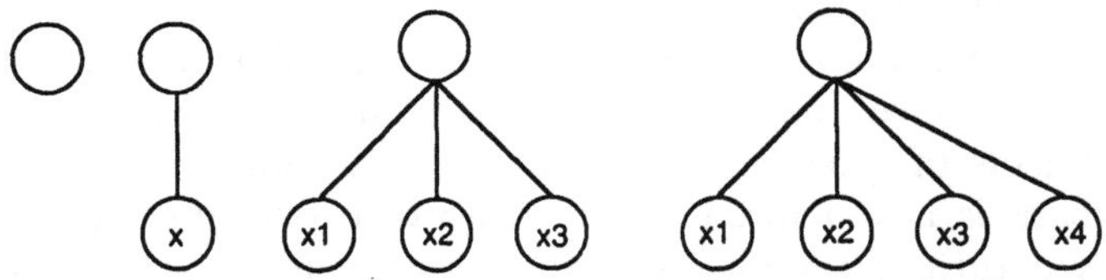

Fig. 4.14. Constellations of SUCCESSORS 1,3..* X

Fig. 4.15. Constellations of SUCCESSORS 1 T AND 1 U

Up to now a structured object could have subobjects of only one type; usually it is necessary, however, to combine heterogeneous subobjects. There are three operators AND, XOR, and OR, which serve to build arbitrarily complex successor definitions from the terms introduced so far. Let *A* and *B* denote arbitrary subexpressions. Then, the operators are defined as follows:

- *AND*: A given set of successors N complies with the declaration *A AND B*, if and only if two disjoint subsets N_a and N_b can be found so that both N_a and N_b comply with the declaration *A* and *B*, respectively. Therefore, every constellation of *A AND B* must contain both subobjects demanded by *A*, and subobjects demanded by *B*. Figure 4.15 shows a constellation satisfying such an expression, where *A* has been instantiated to *1 T*, and *B* to *1 U*, *T*, and *U* being type names (and likewise in Figs. 4.16 and 4.17).

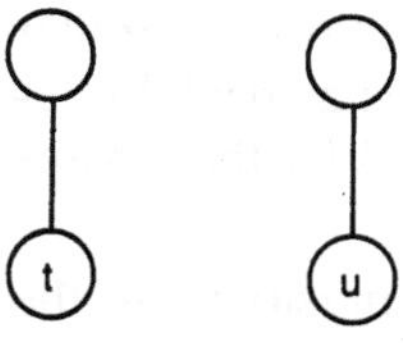

Fig. 4.16. Constellations of SUCCESSORS 1 T XOR 1 U

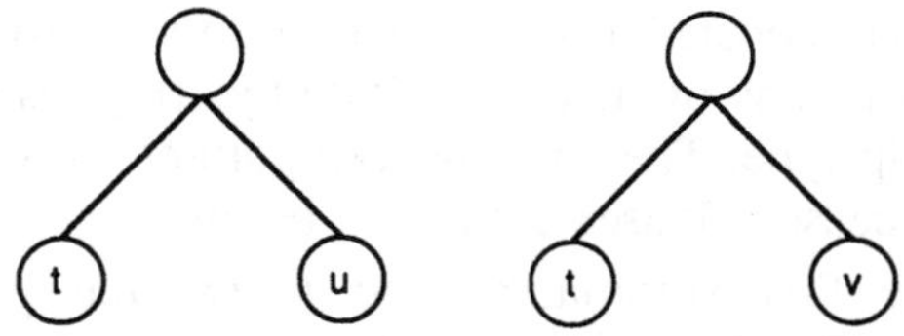

Fig. 4.17. Constellations of SUCCESSORS 1 T AND (1 U XOR 1 V)

- *XOR*: A given set of successors N complies with the declaration *A XOR B*, if and only if it complies either with the declaration *A* or with the declaration *B*. The purpose of XOR is to allow for alternatives (Fig. 4.16). As the last example in Fig. 4.17 shows, a distributivity law holds between AND and XOR: *A AND (B XOR C)* is equivalent to *(A AND B) XOR (A AND C)*.
- *OR*: A given set of successors N complies with the declaration *A OR B*, if and only if it complies either with the declaration *A* or with the declaration *B*, or with the declaration *A AND B*.

As opposed to the successors definition, the predecessors of an object are not defined explicitly. The fact that a certain type appears within the successors clause characterizes objects of this type as potential predecessors of objects of the type indicated.

4.2.2.3 Completeness of Objects

An important requirement in design applications is to know when a design object has reached a certain state of consistency. Although this question can only be answered in many cases by the application itself (e.g., by a circuit simulation), there are some relatively inexpensive checks as well. Checks for conformance with the schema definition belong to this class. This means not only things like value integrity, but the semantic expressiveness of the schema definition language presented above also allows for monitoring a completeness criterion, which will be introduced now.

In the three previous sections, only objects conforming with their type definition were under consideration. But during the design process objects are generated incrementally and, consequently, may be incomplete for a long time. This is taken into consideration by introducing the object status "design-complete" or just "complete". A structured object is called *design-complete* if it forms a constellation conforming to the type definition. A structured object is called *transitive design-complete* if its root and all its subobjects are complete. Now objects can be built successively; the completeness status is controlled by the system and can be obtained at any time.

4.2.2.4 Standard Relationships in Software Engineering

In this section we consider some kinds of relationships which are very typical in the area of SE. They can be represented in PRODAT by fixing certain admissible constellations of successors in the object type definition. Similar combinations of successor (subobject) types can be classified as a special relationship type. The four most important ones in SE are described by the classes of successor clauses explained below.

We first summarize and repeat the predicate logic expresions in the SUCCESSORS clauses and their meaning (semantic) for the term *completeness*:

- *A AND B*: There must be objects of both types, type 'A' and type 'B'.
- *A OR B*: There must be either objects of type 'A', or objects of type 'B', or objects of both types (non-exclusive).
- *A XOR B*: There must be either objects of type 'A', or objects of type 'B' (exclusive).

Let 'k' be a cardinality. We write the following abbreviations for standard relationships. The database system will interpret them as described above. If there is no cardinality given, one may generate as many objects of this type as one wishes.

- For k_1 A_1 AND ... AND k_nA_n, write K (k_1A_1, ..., k_nA_n), to define the *has_component* relationship.
- For 1...k_1A_1 OR ... OR 1...k_nA_n, write A (k_1A_1, ..., k_nA_n), to define the *has_alternative* relationship.
- For 1 A_1 AND ... AND 1 A_n, write R (A_1, ...,A_n), to define the *has_representation* relationship.
- For A_1 OR ... OR A_n, write E (A_1, ...,A_n , to define the *has_element* relationship, which is the most common one.

For each kind of relationship or successor clause, there are different constraints which define the state *complete* for objects described with such a structure. This is explained informally by the following definitions:

- *has_component*: With the *has_component* relationship the user can describe the decomposition of an object and the subobjects that are the parts of this object. The object is complete if all components, i.e., all subobjects, exist (on the instance level).

- *has_alternative*: With the *has_alternative* relationship the user can describe several alternative instances (or realizations) of an object (or problem). The completeness rule for the successor clause reads: At least one subobject of any stated type must exist. Then the state *complete* is reached for the object considered.

- *has_representation*: Each object may exist in different representations to describe the different abstraction levels of a problem. Consequently, we need a relationship between a superobject and its representations. By the *has_representation* relationship the user closely interconnects these representation objects. The corresponding successor clause represents the completeness rule that the described object is complete if exactly one subobject of each stated type (representation type) exists.

– *has_element*: If there exists only a loose connection between objects which are collected in a library or a similar organization, we call them elements. The library may be represented in the database by a structured object, and the elements of the library are connected as subobjects of the latter by the *has_element* relationship. This relationship always fits, since there is no condition about the subobjects in the type definition. Therefore the completeness rule reads: Even if no subobject exists, the object is complete.

4.2.2.5 Versions

As the states of a system's evolution must be documented, it is not desirable – especially in large systems – to overwrite an object when changing it. Various reasons can be given. While changing a module (or a processor, a device driver, etc.), the system to which the module belongs must continue to work. So the old and the new module have to coexist. Another reason is that a programmer might want to work with the state an object was in two weeks ago. This is, of course, only possible if this state is preserved.

As a result of these considerations, the need for a version concept arises. The main aspect mentioned above is the development history (see Sect. 3.2) of the objects. This is the way versions are used in PRODAT. An object is said to be a version of another object, if it is derived from this object. So creating a new version means copying an object, storing the development connection, and leaving the "original object" unchanged, so that other users can carry on working with the original.

There is yet another aspect of versions. Often there is the need to develop objects in parallel (so-called variants), and decide later which of them solves the problem best. So, to support this aspect, the development of versions is not necessarily linear (in this case, the versions are usually called revisions), but may be tree-like, if the versions are seen as nodes and the *is_derived_from* relations as edges. Note that there is no way to merge versions, i.e., to override the tree scheme. The resulting derivation graph is called the *version graph*.

Let us now give a more precise characterization of the terms version, origin, and version graph.

- Every version is an object (and can be treated as such).
- Every object developed from another object by versioning is a version.
- Every object from which a version is derived (the version's *origin*) is also a version.
- All versions derived from one object (possibly over a number of steps) establish one version tree.
- Versions can be generated at any time.
- An object which is the origin of some version must not be changed any more.
- A *version number* is assigned to every version to represent the chronological ordering of derivation. The version number will be written as a suffix to the object name (e.g., program.4 and program.6 are versions of the object program).

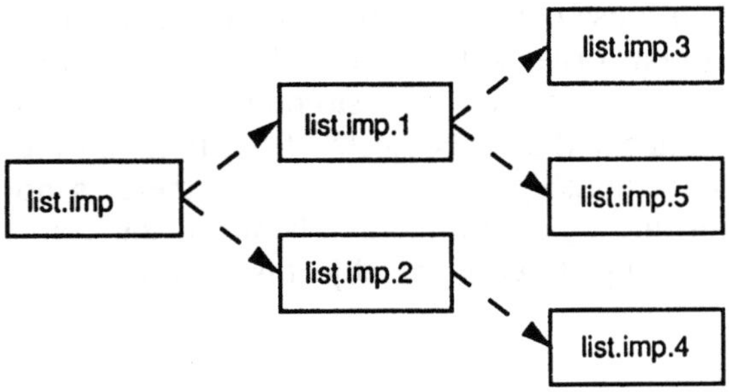

Fig. 4.18. Version tree

An example will show the version tree of an object evolving during its life cycle (Fig. 4.18). The example object is the implementation *list.imp* as part of the structured object *finset* shown in Fig. 4.11. *List.imp.1* is changed several times. First, a data structure is changed, yielding version *list.imp.1*. Then, the variant *list.imp.2* is created to extend the functionality. The version *list.imp.3* is the result of an error correction in *list.imp.1*, while version 5 is obtained by replacing a procedure. The reason for version *list.imp.4* is the extension of checking the preconditions.

The example shows that the reasons for creating a new version are numerous. Only the designer can decide if, for example, revisions are appropriate (linear version sequence), or if variants are to be used (version tree). A basic version mechanism must support both.

In the same way as all other objects, versions can be used as subobjects of structured objects. However, there is one restriction. The user has to decide which version is part of his structured object. It is not possible to say that all objects of the version tree are part of the structured object and choose a particular version later on. The user must instead connect one single version from the tree to his structured object.

It should be noted that versions of structured objects can be derived as well. Although it is possible to build a version of every simple object involved and manually re-establish the structure, it is much more convenient xto do the whole process in one step.

4.2.2.6 Configurations

Systems supporting the whole system life cycle need mechanisms to deal with the components of the deliverable systems. This means choosing all the parts establishing the complete system as well as supporting the customers in the course of maintenance. A configuration management system is usually considered helpful for this purpose.

Structured objects mostly consist of more than just the parts of a system a customer will receive. A good example are source texts which remain with the developer. They are, therefore, part of the structured object describing the project, but not of the configuration object describing the delivery. So the configuration will contain only a selection of all parts available.

In the delivery and maintenance phase of a project, organizational information related to the configurations arises, for instance customer names and dates of expedition. These data are best attached to a configuration by defining additional, configuration-specific attributes.

In PRODAT, configurations have – like any other object – a type. They consist of two main features. First, a configuration instance contains an "inventory" of objects which must be compiled according to the selection rules in the corresponding type definition. Both number and type of objects which may take part in a configuration object are described in the SELECTION clause. Second, additional attributes can be specified which allow for additional information characterizing the configuration.

It is the task of a configuration management tool to fill in more semantics into this basic notion. One can think of tools which automatically select the alternative parts of a structured object considering various conditions like runtime behaviour, main memory requirements, security aspects, price, etc.

4.2.3 A Tool Using PRODAT – The Object Editor

The PRODAT Object Editor (POE) is a graphics tool for inspecting and manipulating the objects of the database interactively. It is composed of two parts, the graphics part and the database part.

The graphics part manages the dialogue between the user and POE. Multiwindowing is used for user communication and displaying structured objects. A structured object is represented as a directed graph. Consequently, the relations between objects are represented by edges between these objects. A simple object is represented by a rectangle with its name inside. We give a short introduction to the principal terminal display.

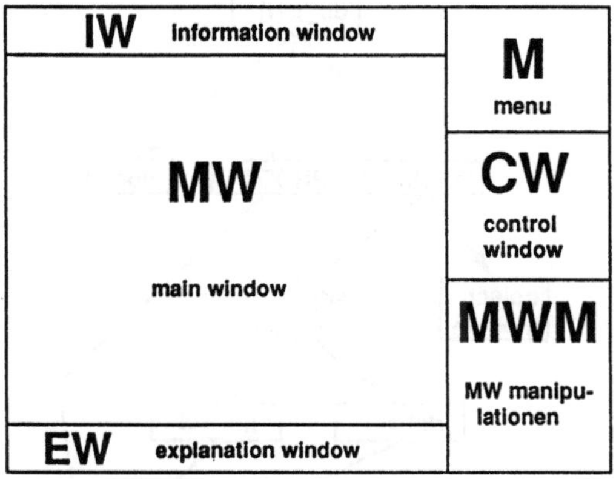

Fig. 4.19. General screen display

Figure 4.19 gives an overview of the terminal display. The display is divided into several windows. The main window (MW) is used for displaying and graphical identification from PRODAT components. Its content can be mainpulated by the main window mainpulation unit (MWM) in the lower right-hand corner. The MWM is composed of several arrows. Each arrow represents one manipulation direction of the main window. It also contains a zooming unit which increases and decreases the size of displayed units in the main window. The information window (IW) contains information for the user, e.g., database name, environment. The explanation window (EW) explains the operations which the user can currently select, or tells the user what input the system is expecting. The menu window (M) is composed of options and operations. These operations can be selected by mouse pick from the user. The last window is the control window (CW). It contains the complete structured object to get an overview. It displays only the structure without xny further information. The user gets more information in the main window, which generally displays only a part of the structured object.

We will not treat the graphics part of POE in detail, because it handles the dialogue to the user, and we are interested in the interaction with the database. The database part supplies data for POE out of the database. It also executes the database operation (e.g., insert object, or delete object) selected by the user. For this manipulation, the database part uses the procedural interface of PRODAT.

The database part will now be described in more detail. For a better understanding, POE will be explained as a structured object of PRODAT (Fig. 4.20). First, we show the type definition, which uses the standard relationships of software engineering described above, and the corresponding terminal session for creating the objects. Afterwards, we show the structured object with all its subobjects.

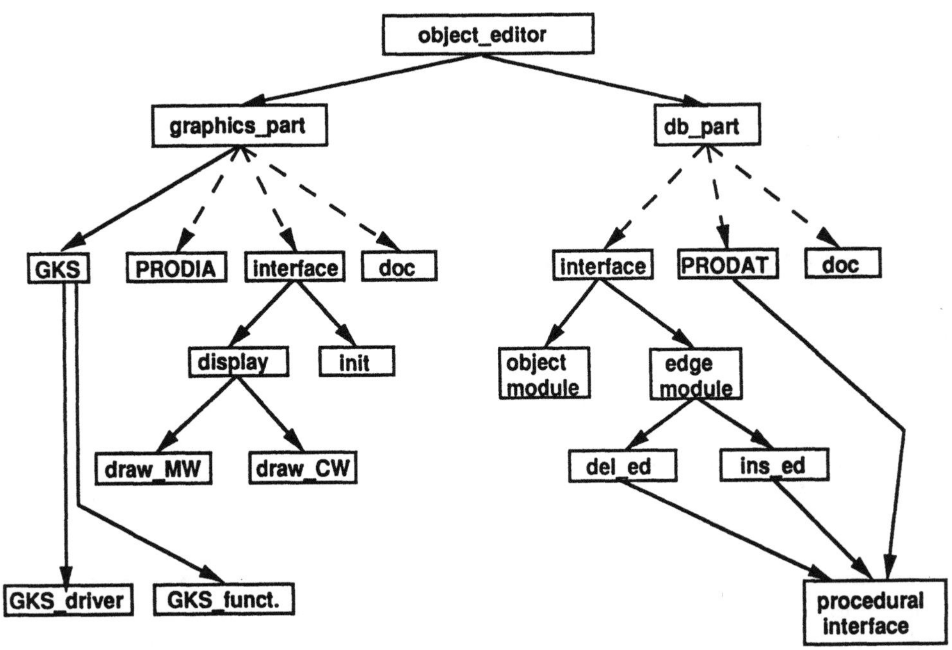

Fig. 4.20. The POE as a structured object

```
TYPE   DOCUMENTATION    STRUCT   [
                                      CONTENTS
                                 ];

TYPE   BASIC_FUNCTIONS  STRUCT   [
                                      Packet CHAR ( 20 ),
                                      Language CHAR ( 20 ),
                                      Version CHAR ( 10 )
                                 ];

TYPE   INTERFACE_FCT    STRUCT   [
                                      Compilation DATE,
                                      Interface_name CHAR ( 20 ),
                                      CONTENTS,
                                      K ( 1 BASIS_FUNCTIONS )
                                 ];

TYPE   MODULE           STRUCT   [
                                      Ok BIT ( 1 ),
                                      Module_name CHAR ( 20 ),
                                      CONTENTS,
                                      E ( INTERFACE_FCT )
                                 ];

TYPE   PACKAGE          STRUCT   [
                                      Name CHAR ( 20 ),
                                      Task CHAR ( 80 ),
                                      CONTENTS,
                                      K ( 1,2 BASIC_FUNCTIONS )
                                 ];

TYPE   INTERFACE        STRUCT   [
                                      POE_part CHAR ( 20 ),
                                      CONTENTS,
                                      E ( MODULE )
                                 ];

TYPE   PART             STRUCT   [
                                      POE_part CHAR ( 20 ),
                                      CONTENTS,
                                      R ( PACKAGE,
                                          DOCUMENTATION,
                                          INTERFACE )
                                 ];

TYPE   PROJECT          STRUCT   [
                                      Author CHAR ( 40 ),
                                      Members CHAR ( 40 ),
                                      Institut CHAR ( 20 ),
                                      CONTENTS,
                                      K ( 2 PART )
                                 ];
```

The user now has two possiblities for creating the structured object. Either he sets up the object interactively with the help of graphics, or he uses (as shown below) the command interpreter. The command interpreter allows manipulation of the database by an alphanumeric interface. There are two functions for creating this object. The function "crob" generates an object and "cred" an edge. The variable "DB" stands for the database name and "key" for the several object keys. The "crob" function needs the following parameters: dbname, objtype, objname, fatherkey, objkey. Function "cred" only needs the two keys of objects, which should be related. Furthermore, the objectkey, which is an output parameter, is only named as a variable (keynr), if it is needed again as a fatherkey; otherwise it is empty.

```
crob  DB  ,PROJECT           ,object_editor          ,0      ,key1   ;
crob  DB  ,PART              ,graphics_part          ,key    ,key2   ;
crob  DB  ,PART              ,db_part                ,key1   ,key3   ;
crob  DB  ,PACKAGE           ,GKS                    ,key2   ,key4   ;
crob  DB  ,PACKAGE           ,PRODIA                 ,key2   ,       ;
crob  DB  ,INTERFACE         ,interface              ,key2   ,key5   ;
crob  DB  ,DOCUMENTATION     ,doc                    ,key2   ,       ;
crob  DB  ,BASIC_FUNCTIONS   ,GKS_driver             ,key4   ,       ;
crob  DB  ,BASIC_FUNCTIONS   ,GKS_funct.             ,key4   ,key14  ;
crob  DB  ,MODULE            ,display                ,key5   ,key6   ;
crob  DB  ,MODULE            ,init                   ,key5   ,       ;
crob  DB  ,INTERFACE_FCT     ,draw_MW                ,key6   ,key7   ;
crob  DB  ,INTERFACE_FCT     ,draw_CW                ,key6   ,key8   ;
crob  DB  ,INTERFACE         ,interface              ,key3   ,key9   ;
crob  DB  ,MODULE            ,object_module          ,key9   ,       ;
crob  DB  ,MODULE            ,edge_module            ,key9   ,key10  ;
crob  DB  ,PACKAGE           ,PRODAT                 ,key3   ,key11  ;
crob  DB  ,DOCUMENTATION     ,doc                    ,key3   ,       ;
crob  DB  ,INTERFACE_FCT     ,del_ed                 ,key10  ,key12  ;
crob  DB  ,INTERFACE_FCT     ,ins_ed                 ,key10  ,key13  ;
crob  DB  ,BASIC_FUNCTIONS   ,procedural_interface   ,key11  ,key15  ;
cred  key7   ,key14  ;
cred  key8   ,key14  ;
cred  key12  ,key15  ;
cred  key13  ,key15  ;
```

As one can see above, all objects are created by the crob operation, whereby they are connected directly to their father object. If another hierarchical relationship is needed between an object and a superobject, different to the one identified in the crob operation, one has to use the cred operation. By using the cred operation, shared subobjects are generated. The session creates the structured object as shown in Fig. 4.20. Figure 4.20 shows the PRODAT Object Editor as a structured object. The picture only displays the global interrelations, by showing the structure in a very simple form. The relations between the graphics and database interface are missing.

The POE is composed of two *components*, the graphics part and the database part. The graphics part is *represented* by GKS, PRODIA, interface, and documentation. GKS contains two *components*, the GKS driver and the GKS functions. In this object, PRODIA and the documentation are not detailed any further. The interface of the graphic part is a *library*, which contains modules. As examples, the display and the initialization modules are shown. Each module is a *sublibrary*, so the display module includes the functions draw_main_window and draw_control_window. Each function uses *components* from the GKS functions.

The *representations* for the database part are an interface, the PRODAT part, and documentation. The details of the documentation of the database part are omitted. PRODAT is only displayed with one *component*, the procedural interface. The database interface is a *library*, which contains simplified views of the two *sublibraries* object and edge module. The edge module is further divided into two functions, delete_edge and insert_edge, which use *components* of the procedural interface of PRODAT.

The modules of the database interface can be separated generally into two classes. One class represents the modules for manipulation of the database. The second class of modules contains all inquire functions and display functions.

The modules belonging to the first class are the object, attribute, content, and edge modules. The object module creates or deletes an object. Additionally, it renames simple objects. The attribute module contains functions for manipulating the attributes of simple objects. This module also belongs to the second class of modules. The content module manipulates the content of an object. For creating or deleting an edge, the edge module uses the procedural interface functions "pom_cred" (cred = create edge) and "pom_deled" (deled = delete edge). The edge module, e.g., performs the most important task of the interface: It transfers the data structures between the graphics and the database parts. The first class of modules always executes actions in the database, which change the state of it.

The second class of modules for inquiring and displaying information includes modules for keeping database and starting-object names ready, the module for setting up the graph, the attribute module, and the object-state module. The module for database and starting-object names delivers the names in alphabetical order. The most complex module is for setting up the graph of a structured object. Because of the dynamic data structure it uses recursive algorithms to get the objects. It uses the procedural interface functions "pom_gaparob" (gaparob = get all parent objects) and "pom_gachildob" (gachildob = get all child objects) to get the keys of the parents and children of an object. Afterwards, it prepares the data structure for the graphic display. The attribute module displays the attributes of an object. The module of object states delivers the dates of a simple object like creation date, date of last change, etc.

In summary, this example has introduced the POE as a tool which is able to display and manipulate the complex structures of databases and objects in a comfortable way.

4.2.4 PRODAT as Interface to DAMOKLES in Software Engineering

In this section, we compare the DAMOKLES data model with the PRODAT approach by mapping a PRODAT schema into one of DAMOKLES. For this case study, DAMOKLES may serve as a database kernel system for engineering applications, and PRODAT may take the part of an application-oriented software engineering database interface. Considering only the modeling concepts and mechanisms for complex objects, the main difficulties in modeling data structures in this area are shown.

The first step in using DAMOKLES to realize a PRODAT database is to map each feature of its object type definition into one of the kernel data model. Therefore, we have to consider the TDL expressions and the DAMOKLES DDL (DDDL). An example of their basic mechanisms for structured object definition is shown in the following:

```
typedef   Object {
          NAME c_name ;
          BOOLEAN released ;
          NAME author ;
          . . .
          contents ;
          successors 1..10 PORTS
                 and INSTANCES
                 and 1 REF
                 and 1..10 CONNECT ;
          } CELL
```

Object definition in PRODAT TDL

```
OBJECT   TYPE cell
         ATTRIBUTES
         c_name : name ;
         released : boolean ;
         author name ;
         description : longfield ;
         . . .
         STRUCTURE IS
         ports, instance, ref, connect
         AT MOST ONCE (ref)
         AT LEAST ONCE (ref)
         END  cell
```

Object definition in DAMOKLES DDL

Some parts of the object definition are very easy to map, such as the object type name, which is given to a type in the same way in both systems. Also, the attribute definition part for objects seems to be very similiar. But in this case, we have to be careful with the attribute domain types. Certainly, we can find a mapping between basic types like integer, character, boolean, and real. For all of these, a comparison between the internal representations is necessary to ensure a correct representation of PRODAT types in DDDL. The next level of attribute domain types are user-defined attribute types like enumeration types, arrays, strings, etc. Both data models allow the definition of such types. Because they are more high-level constructions, they are not dependent on internal physical representations like the basic types. Therefore, it is possible to map one user-defined TDL type into another one in DDDL, especially for each PRODAT user type, because DAMOKLES is very powerful in the area of type definition capabilities. For further detailed information of the definition languages, see the reading list at the end of this chapter.

Another part of an object description is the contents or long-field definitions. In PRODAT, you can define exactly one contents for an object, whereas in DAMOKLES each object can have several such internal byte strings, called long-fields. This is different in both definition languages: In the TDL you only can switch on if there should exist a contents for an object. In the DDDL you can define more than one contents, and they are named. In the latter, they are treated like special attributes of type long-field. In summary, it is easy to represent contents in DAMOKLES.

The next part of the object type definition is the description of the possible structure of an object. In this part, we see the most differences between the PRODAT object model and that of DAMOKLES. For structured objects, the representation is restricted to the composition of complex objects with the standard relationships explained above in Sect. 4.2.2.4. On the one hand, they describe the most common object hierachies; on the other hand, a total mapping of successor predicates into a general entity relationship model with special structuring capabilities is a really theoretical approach, which will not be considered here. The goal of this section is to give an idea of some consistency constraints for the structure of complex objects, which could not be formulated with basic features of classical data models like the relational or network model. The description of the allowed structure of an object is generally done by the successor clause (PRODAT) or the structure definition (DAMOKLES). In PRODAT, the successors (subobjects) which compose a structured object are defined at the type level by the successor clause. Each successor clause (exactly one per object) includes a certain consistency rule about the composition of the structured objects of the object type concerned: How should subobjects be composed into a superobject? These consistency rules are closely related to an object type. Although this kind of consistency definition is static, it allows consistency checking during the dynamic structuring process. The system is able to inform the user about the state of completeness of an object (see Sect. 4.2.2). This is done by several system attributes for each object which describe the status of this object. On the other hand, some restrictions on the structure are useful, and they allow basic consistency mechanisms, which are model-inherent in the case of PRODAT. In DAMOKLES, there is no integrated mechanism to express conditions about the structure of objects. This general independence in structuring an object is comparable to the *has_element* standard relationship, which can be directly represented in DDDL through the structure definition.

has_element:
Object TYPE TE (SUCCESSORS A or B or C);
is the same as,
Object TYPE TE SUCCESSORS E (A,B,C);
and is mapped to :
Object TYPE TE STRUCTURE is A,B,C END TE;

In the structure definition of DAMOKLES, the listed subtypes could either be an object type or a relationship type. The relationship types are not used for modeling PRODAT, because these do not compose objects by explicitly defining relationships. They are implicitly defined by the successor clause, and there is no relation between the subobjects except their common superobject (predecessor). But you can define a cardinality (Sects. 4.2.2 and 2.5) for the relationship between this superobject and the subobjects. Additionally, there is a possibility for adding semantics to this special relationship by defining conditions about the successors, which can coexist. This kind of semantics comprises special integrity rules, which have to be inserted in the database system to provide behavior similiar to PRODAT.

For that purpose, DAMOKLES will be extended by a consistency control mechanism, which ensures checking inside (the database system) instead of checking outside in application programs. In engineering applications, a generalized consistency control mechanism as a separate component of the database system is more powerful than a specialized one. Using such a mechanism, we first have to formulate the consistency rules for each standard relationship explicity and informally, as we have done in Sect. 4.2.2.4. These rules are checked as completeness rules automatically, when using the PRODAT operations create_object, create_edge, delete_object, and delete_edge. For DAMOKLES we need a mechanism which allows the definition of such rules or constraints, and the connection of such constraints to a certain event like the creation of a so-called edge. Furthermore, the mechanism must include the definition and execution of reactions on checks. Such a mechanism is the Event/Trigger Mechanism (ETM), which is based on three components:

- events,
- actions, and
- triggers.

An event signals a situation to which reactions may be necessary. An action is an arbitrary program module to be executed. It may be defined in some high-level language, including database operations. A trigger T associates an event E with an action A, meaning that the raising of event E implies the execution of action A. Triggers are specified by

TRIGGER <T_id> = ON <E_id>
DO <action> (parameters).

You can activate and deactivate triggers dynamically by a special operator, and also delete triggers, events, and actions.

The definition of standard events for the object and edge manipulation operations is necessary to associate such events with the checking of the constraints and the starting of actions. The standard events are:

For *create_object*: EVENT $crob,
For *create_edge*: EVENT $cred,
For *delete_object*: EVENT $delob, and
For *delete_edge*: EVENT $deled.

The database system has to manage four sets of types to partition all defined object types into the disjunctive sets TK, TA, TR, TE, whereby:

- TK contains all types with standard relationship *has_component*,
- TA contains all types with standard relationship *has_alternative*,
- TR contains all types with standard relationship *has_representation*, and
- TE contains all types with standard relationship *has_element*.

When creating a new object by *create_object*, the system has to check certain constraints in order to guarantee completeness rules. For each kind of standard constellation in the successors clause (standard relationship), we first require checking of the cardinality for a given subobject type (is it allowed to insert a new subobject of the given type). In the case of a successful cardinality check, a new event is raised which triggers the checking of the constellation (coexistence of other subobjects of different types). Together, both checks – the cardinality and the constellation tests – ensure the existence of subobjects which fulfill the completeness rules (as consistent database states). As an example for the first step – the cardinality check – we show here the constraint which defines the conditions for representations. The following informal definition is given in a procedural PASCAL-like syntax:

```
CONSTRAINT rep_unique (objkey, subobjkey: keytype) =

(* The constraint controls the cardinality condition *)
(* (unique existence of subobjects of a certain type) *)
(* for subobjects of an object of a type which is contained in TR *)
(* Reader: compare the definitions of representations in section 4.2.2.4  *)

    VAR childkeylist: key_type_list;
      child : key_type;
      objtype, childtype, subobjtype : type;

    BEGIN
        gettype (objkey, objtype); IF objtype   TR THEN BEGIN
            getallchildobj (objkey, childkeylist) IF  child  childkeylist
                where gettype (child, childtype) = gettype (subobjkey, subobjtype)
            THEN RETURN (false) ELSE RETURN (true)
        END ELSE RETURN (true);
    END;
```

The consistency checks for representations are triggered, for example, in the case of create_object, by raising the standard event through the following triggers:

```
TRIGGER rep_cardinality    =    ON $crob (objkey, subobjkey)
                                DO $check.rep_unique (objkey, subobjkey).
```

The constraint for checking the actual constellation of an object is named *rep_constellation*, and will be used in the second trigger. The implementation of that constellation check is not shown here.

```
TRIGGER rep_complete   =   ON $rep_unique.SUCCESS
                           DO $check.rep_constallation (objkey, subobjkey).

TRIGGER rep_reject1    =   ON $rep_unique.FAIL
                           DO $reject.

TRIGGER rep_reject2    =   ON $rep_constellation.FAIL
                           DO $reject.
```

The Event/Trigger Mechanism automatically generates, from a constraint, a check procedure (action) which is called by *$check.constraintname*. Additionally, the system manages two events: the event *$constraintname.FAIL* and the event *$constraintname.SUCCESS*, which can arise automatically as a result of the check. For the representation example, the checking action *$check.rep_unique* induces two events, *$rep_unique.FAIL* and *$rep_-unique.SUCCESS*. If the check for uniqueness was successful (.SUCCESS), another trigger, named *rep_complete*, for checking the constellation is needed; otherwise, (.FAIL) the triggers *rep_reject1* and *rep_reject2* are defined to reject the standard event *create_object*. Rejection of events or database operations is a reaction to guarantee consistent database states.

In summary, the example has shown the flexibility of an external consistency control mechanism which has been used for modeling semantic rules on the basis of data types and structures. These special semantic rules can be model-inherent as well (see PRODAT data model), but generally most engineering applications need a capability for adding semantics to the objects and relationships in the form of consistency constraints. Therefore, a powerful control mechanism should support database consistency.

4.3 Further Reading

The following papers cover some of the material in the context of EDIF and E.I.S (Sect. 4.1):

Dadam P., et.al.: *A DBMS Prototype to Support Extended NF^2 Relations: An Integrated View on Flat Tables and Hierarchies.* Proc. Int. Conf. on Management of Data, Washington D.C., May 1986

EDIF Steering Committee: EDIF Electronic Design Interchange Format Version 200. Electronic Industries Association, Washington D.C., 1987

Klahold P., Köhler D., Ungerer M., Wilkes W.: *Darstellung von Design-Objekten in relationalen Datenbanken.* Tagungsband zum 3. E.I.S.-Workshop, GMD-Studien Nr. 126, Bonn 1987 (in german)

Klahold P., Wilkes W., Ungerer M.: *Datenmodelle für VLSI-Entwurfsdatenbanken.* Proc. GI-Fachgespräch Non-Standard-Datenbanken für graphische Anwendungen, März 1988 (in german)

Ungerer M., (Hrsg.): *CAD-Schnittstellen und Datentransferformate im Elektronik-Bereich.* Beiträge zur graphischen Datenverarbeitung, Springer-Verlag 1987 (in German)

Some more information about the topics of chapter 4.2 are given in:

Batz T., Baumann P., Köhler D.: *A Data Model Supporting System Engineering.* Proc. of the IEEE 12th Int. COMPSAC, Chicago, USA, October 1988

Dittrich K.R., Gotthard W., Lockemann P.C.: *DAMOKLES – Database System for Software Engineering Environments.* Lecture Notes Comp. Science 244, Springer, 1987, 353-371

Dittrich K.R., Kotz A.M., Mülle J.A.: *An Event/Trigger Mechanism to Enforce Complex Consistency Constraints in Design Databases.* SIGMOD Record 15 (1986) 3, 22-36

Kotz A.M., Dittrich K.R., Mülle J.A.: *Supporting Semantic Rules by a Generalized Event/ Trigger Mechanism.* Proc. Int. Conf. on Extending Database Technology, Venice, March 1988

MIX
Papier aus verantwortungsvollen Quellen
Paper from responsible sources
FSC® C105338

If you have any concerns about our products,
you can contact us on
ProductSafety@springernature.com

In case Publisher is established outside the EU,
the EU authorized representative is:
Springer Nature Customer Service Center GmbH
Europaplatz 3, 69115 Heidelberg, Germany

Printed by Libri Plureos GmbH
in Hamburg, Germany